T0324173

Springer Theses

Recognizing Outstanding Ph.D. Research

Aims and Scope

The series "Springer Theses" brings together a selection of the very best Ph.D. theses from around the world and across the physical sciences. Nominated and endorsed by two recognized specialists, each published volume has been selected for its scientific excellence and the high impact of its contents for the pertinent field of research. For greater accessibility to non-specialists, the published versions include an extended introduction, as well as a foreword by the student's supervisor explaining the special relevance of the work for the field. As a whole, the series will provide a valuable resource both for newcomers to the research fields described, and for other scientists seeking detailed background information on special questions. Finally, it provides an accredited documentation of the valuable contributions made by today's younger generation of scientists.

Theses are accepted into the series by invited nomination only and must fulfill all of the following criteria

- They must be written in good English.
- The topic should fall within the confines of Chemistry, Physics, Earth Sciences, Engineering and related interdisciplinary fields such as Materials, Nanoscience, Chemical Engineering, Complex Systems and Biophysics.
- The work reported in the thesis must represent a significant scientific advance.
- If the thesis includes previously published material, permission to reproduce this must be gained from the respective copyright holder.
- They must have been examined and passed during the 12 months prior to nomination.
- Each thesis should include a foreword by the supervisor outlining the significance of its content.
- The theses should have a clearly defined structure including an introduction accessible to scientists not expert in that particular field.

More information about this series at http://www.springer.com/series/8790

Wenbo Chu

State Estimation and Coordinated Control for Distributed Electric Vehicles

Doctoral Thesis accepted by
Tsinghua University, Beijing, China

 Springer

Author
Dr. Wenbo Chu
Department of Automotive Engineering
Tsinghua University
Beijing
China

Supervisors
Prof. Huei Peng
Department of Automotive Engineering
Tsinghua University
Beijing, China

and

Department of Mechanical Engineering
University of Michigan
Ann Arbor, USA

Prof. Keqiang Li
Department of Automotive Engineering
Tsinghua University
Beijing
China

Prof. Yugong Luo
Department of Automotive Engineering
Tsinghua University
Beijing
China

ISSN 2190-5053   ISSN 2190-5061   (electronic)
Springer Theses
ISBN 978-3-662-48706-8   ISBN 978-3-662-48708-2   (eBook)
DOI 10.1007/978-3-662-48708-2

Library of Congress Control Number: 2015953789

Springer Heidelberg New York Dordrecht London

Printed on acid-free paper

Springer-Verlag GmbH Berlin Heidelberg is part of Springer Science+Business Media
(www.springer.com)

To Jiani

Supervisors' Foreword

The pursuit for safer and more efficient vehicles is a challenging task that has attracted many of the brightest engineers and scientists. With more sensors and active controls available on ground vehicles, they have become safer than ever. Recently, driver assistance functions are widely deployed, and the prospect of driverless operations is closer to reality than ever. It is widely recognized that there are many challenges in sensing, perception, decision, and control before fully driverless vehicles can be safely used. In particular, the ability to perceive the driving environment accurately and reliably seems to be a critical link that requires a lot more work. This book addresses several related questions at a fundamental level.

The two sets of key problems addressed in this book are state estimation and coordinated control. To control the motion of a vehicle, it is important to know the vehicle parameters and states. Some of them are hard to measure or estimate accurately. Under dynamic driving conditions, the environment keeps changing, making the estimation problems even harder to solve. Chapter 3 of this book tackles four of the most challenging problems, and the developed methods make it possible to have more accurate information regarding the vehicle states, tire forces, and road load. This is an important first step to achieve better vehicle motion controls, e.g., to crash avoidance with pedestrians or other vehicles.

Chapter 4 of this book shows that when more accurate vehicle states and forces are available, coordinated control makes it possible to ensure better desired vehicle motions and ensure better robustness under uncertainties in the driving environment. The technical challenges of the estimation and control problems were reviewed briefly in Chap. 2, which explains the key ideas and their importance. The book ends with comprehensive simulation and experimental studies, to demonstrate how the proposed estimation and control methods work under a variety of realistic conditions.

Active safety systems have been widely deployed and have demonstrated their benefits to the Society. Automated vehicle function have started to appear on production vehicles, making them more convenient, comfortable, and easier to operate. This book makes a fine contribution on model-based estimation and control methods, and lays the foundation for more precise control of vehicle motions. This book is technically deep and rigorous on these important problems, and is one of the most comprehensive publications I have seen in recent years.

Ann Arbor   Prof. Huei Peng
August 2015   Prof. Keqiang Li
                                                                            Prof. Yugong Luo

Acknowledgments

I owe thanks to Tsinghua University, which is the spiritual homeland for all THU graduates. Self-discipline and social commitment are values of Tsinghua University. A decade ago, I came to the fantastic garden with dreams, immersing into wisdom and guide from academic masters, encouraged by excellent peers. I studied hard with heart and soul, making endless efforts for study. Memories in Tsinghua University become part of my life. I learned and knew friends here, finding myself and my dreams.

It was appreciated that my supervisor, Professors Huei Peng, Keqiang Li, and Yugong Luo taught me deeply and widely. They offered important experiences and knowledge to me from macro-direction to details as typography. When I met with obstacles as bottleneck of the thesis, they contacted me pro-actively. It was appreciated that in process of research and thesis compilation, they helped and supported me a lot. Without such help and supports, I could hardly do it so strictly and solidly. Their wide knowledge, intelligence, and diligence benefit me for lifelong. Thank you!

I owe thanks to Professor Xiaomin Lian, Jianqiang Wang, Diange Yang, Sifa Zheng, and Shengbo Li for good proposals! I thank Jingjing Fan, Jinhui Zhang, Yanchao Wang as the seniors from State Key Laboratory of Automotive Safety and Energy of Tsinghua University and Tao Chen, Yifan Dai, Yunwu Han, Feng Zhao, Zhaosheng Zhang, Ruina Dang, Shichun Yi, Jian Luo, Qingyun Jiang, Guopeng Luo, Xiaodan Fu, Shuwei Zhang, Kun Xiang, Ming Zhou, Long Chen, Kun Cao, Tao Zhu, Yong Xiang, Shuang Wan, Zhaobo Qin, and other students, who supported me so much. They are the first readers of my works, also the readers with the most patience and faith. Their ideas lighted my fervor, seeping into my works. They play an important role in completion of the whole book.

It is the same as other books, based on previous researchers' works which offer plenty of references, remarks, and notes. That is the best gift for me, as a scientific researcher. Thank you!

Years have passed by, and my wife, Jiani, is always supporting me with faith, patience, and love. Usually her companion could make me devoted to research and innovate with heart and soul. I am grateful to my parents, who offer me endless energy in this scientific way. Gains originate from my family and thank you for companionship and self-giving dedication. Best regard and wishes to all!

Contents

Acronyms

A	Vehicle windward area
a_x	Longitudinal acceleration
$a_{x,m}$	Measured value of longitudinal acceleration sensor
$a_{x,b}$	Static deviation of longitudinal acceleration sensor
a_y	Lateral acceleration
$a_{y,m}$	Measured value of lateral acceleration sensor
$a_{y,b}$	Static deviation of lateral acceleration sensor
a_ω	Hub acceleration
b_f	Front wheel tread
b_r	Rear wheel tread
C_d	Drag coefficient
C_f	Front axle cornering stiffness
C_r	Rear axle cornering stiffness
C_i	Tire cornering stiffness
C_{roll}	Body roll damping
F_x	Vehicle longitudinal driving force
$F_{x,d}$	Desired longitudinal driving force
F_y	Vehicle Lateral force
$F_{i,X}$	Longitudinal force of the i wheel under vehicle coordinate system
$F_{i,Y}$	Lateral force of the i wheel under vehicle coordinate system
$F_{i,x}$	Longitudinal force of the i wheel under tire coordinate system
$F_{i,y}$	Lateral force of the i wheel under tire coordinate system
$\overline{F}_{i,y}$	Quasi-static Lateral force of the i wheel
$F_{i,z}$	Vertical force of the i wheel
$F_{i,max}$	Maximum longitudinal force of the i wheel
f	Road surface rolling resistance coefficient
f_L	Cut-off frequency
G_{ff}	Feedforward direct yaw moment gain
G_{fb}	Feedback direct yaw moment gain
$G_{x\alpha}$	Impact factor of sideslip on longitudinal force

$G_{y\kappa}$	Impact factor of sideslip on Lateral force
g	Gravitational acceleration
h_g	Height of vehicle center of mass
h_r	Height of vehicle roll center
h_s	Height of vehicle center of mass to roll center
I_z	Vehicle yaw rotational inertia
I_x	Vehicle roll rotational inertia
K_P	Proportional error factor
K_{a_P}	Accelerator pedal opening amplification coefficient
K_{roll}	Body roll angle stiffness
k_{a_x}	Acceleration weight coefficient
k_i	Wheel velocity weight coefficient
$k_{y,t}$	Roll stiffness of tire system
L	Lease square gain
l	Vehicle wheel base
l_f	Distance from front axle to center of mass
l_r	Distance from rear axle to center of mass
M_o	Principal moment
M	Yaw resultant moment
M_z	Direct yaw moment
$M_{z,d}$	Desired direct yaw moment
M_{ff}	Feedforward direct yaw moment
M_{fb}	Feedback direct yaw moment
m	Full vehicle mass
P	Error covariance update
Q	Process noise covariance matrix
q	Process noise
q	Side slip angle weight coefficient
r_{min}	Lower limit of motor moment change slope
r_{max}	Upper limit of motor moment change slope
R	Measurement noise covariance matrix
R	Tire roll radius
r	Measurement noise
$T_{i,a}$	Actual driving moment of the i wheel
$T_{i,d}$	Expectation driving moment of the i wheel
$T_{i,u}$	Driving moment command of the i wheel
T_{nl}	Maximum motor moment under normal load
T_{ol}	Maximum motor moment under overload
T_{fc}	Maximum motor moment under failure
T_{TCS}	Drive anti-skid controller driving force
t	Time
u	Input vector
$v_{i,x}$	Velocity of wheel center in direction of tire rolling
v_{GPS}	Vehicle velocity measured by GPS

$v_{i,\omega}$	Hub velocity
v_{x0}	Initial longitudinal braking velocity
v_x	Longitudinal vehicle velocity
v_y	Lateral vehicle velocity
$v_{i,x}$	Longitudinal velocity of wheel center under tire coordinate system
$v_{i,y}$	Lateral velocity of wheel center under tire coordinate system
α	Tire slip angle
α_P	Accelerator pedal opening
β	Side slip angle
$\beta_{0,n}$	Steady state side slip angle without DYC
$\beta_{0,w}$	Steady state side slip angle with DYC
β_d	Target side slip angle
γ	Yaw rate
$\gamma_{0,n}$	Steady state Yaw rate without DYC
$\gamma_{0,w}$	Steady state Yaw rate with DYC
γ_b	Static deviation of Yaw rate sensor
γ_d	Target Yaw rate
γ_m	Measured value of Yaw rate sensor
ΔT	Time step
δ	Front wheel rotational angle
ε	Drive/brake efficiency
ε	Tire load rate
θ	Road surface gradient
κ	Tire slip rate
κ_{opt}	Optimal tire slip rate
λ	Forgetting factor
μ	Road adhesion coefficient
$\mu_{x,p}$	Longitudinal peak adhesion coefficient
$\mu_{y,p}$	Lateral peak adhesion coefficient
ρ	Air density
τ	Time constant
τ_a	Actual system response time constant
τ_d	Expected system response time constant
ϕ	Body roll angle
ω_i	Tire rotate velocity
ψ_V	Vehicle direction angle
ψ_{GPS}	Vehicle course angle
$\circledast_a$	Actual value
$\circledast_b$	Sensor static deviation
$\circledast_d$	Expected value
$\circledast_i$	$i \in \{1, 2, 3, 4\}$, successively represents the left front wheel, right front wheel, left rear wheel and right rear wheel
$\circledast_k$	Variable at time k
$\circledast_{k|k-1}$	Variable at time k recurred by time $k-1$

$\circledast_m$	Sensor measured value
$\circledast_{max}$	Maximum value
$\circledast_{TCS}$	TCS system
$\circledast_x$	Longitudinal
$\circledast_y$	Lateral
$\circledast_z$	Vertical
$\circledast_R$	Coordinate system with origin at wheel center and coordinate axes parallel to those of vehicle coordinate system
$\circledast_V$	Vehicle coordinate system
$\circledast_W$	Tire coordinate system
$\circledast_f$	Front wheel and axle
$\circledast_r$	Rear wheel and axle
$\hat{\circledast}$	Estimated value
ABS	Anti-lock Braking System
ACC	Adaptive Cruise Control
CoG	Center of Gravity
DYC	Direct Yaw Moment Control
EKF	Extended Kalman Filter
ESP	Electronic Stability Program
GPS	Global Positioning System
HAC	Hill-Start Assist Control
INS	Inertial Sensor
KF	Kalman Filter
LQR	Linear Quadratic Regulator
PF	Particle Filter
RCP	Rapid Control Prototype
RISF	Reliability Indexed Sensor Fusion
RLS	Recursive Least Squares
TCS	Traction Control System
UKF	Unscented Kalman Filter
UPF	Unscented Particle Filter
UT	Unscented Transformation

Chapter 1
Introduction

Abstract In a broad sense, electric automobiles refer to the motor vehicles based on electric drive, which can be divided into electric vehicles, hybrid electric vehicles, fuel cell vehicles and other electric-driven vehicles in terms of energy consumed, and into centralized drive vehicles and distributed drive vehicles in terms of power system layout. As the most common mode for internal combustion engine automobiles, centralized drive was designed based on traditional vehicles; its power acts on the wheels through clutch, transmission, drive shaft, differential mechanism, half axle, etc. This sort of design maintains the compatibility between electric automobiles and traditional internal combustion engine automobiles to the maximal extent, which is mainly adopted by hybrid electric automobiles. However, its disadvantages, including large quantity of drive parts, low drive efficiency and complex control, are showing up themselves due to the restriction of traditional automobile design concept; on the contrary, the advantages of electric vehicles, including less mechanical drive links, short drive chain and flexible layout, are gradually discovered, with the constant deepening of design concept for electric automobiles and progress in developing electric drive system. With distributed drive, such drive parts as clutch, transmission, drive shaft, differential mechanism and half axle are spared and the drive motor can directly be installed in or near the driving wheel. As this brand-new chassis form designed for electric automobiles according to the motor characteristics greatly promotes the transformation of automobile structure, it has become a hot spot in research and design.

1.1 Overview

Compared with centralized electric drive automobiles, distributed electric drive automobiles are endowed with the following advantages. First, response of control execution unit is rapid and accurate. The power system of traditional internal combustion engine automobiles is composed of internal combustion engine, clutch, transmission, differential mechanism, half axle and driving wheel, leading to long drive chain, slow dynamic response (actual delay may be up to 100 ms), as well as large output moment error and low real-time control precision of internal combustion engine;

© Springer-Verlag Berlin Heidelberg 2016
W. Chu, *State Estimation and Coordinated Control for Distributed Electric Vehicles*, Springer Theses, DOI 10.1007/978-3-662-48708-2_1

while distributed electric vehicles features short drive chain and rapid motor response speed, making rapid and accurate control possible. Second, the drive system is efficient and energy-saving. For many drive parts have been spared, the driving efficiency of distributed electric vehicles is greatly enhanced; meanwhile, distributed motor can generate braking force, which may lead to reduction of energy consumption by combining with braking energy recovery system. Third, the full vehicle is compact in structure and modularized in design. The adoption of electric drive wheels simplifies the structure of full vehicle to a great extent, and the abandonment of complex drive devices not only reduces failure rate, but also makes higher the interior space utility rate. Fourth, multiple information units are equipped. As the physical properties of distributed drive motor can be revealed through such state properties as voltage and current, the distributed drive motor is both execution unit and information unit. The information unit can make correct feedback on the current in-wheel velocity and driving moment, providing sound foundation for integration of multiple information units. Last, it is equipped with redundancy configuration of multiple power units. The distributed electric vehicles are equipped with multiple independent and controllable power units, which can realize control functions with relative independence; the redundancy configuration is quite favorable for failure control and coordinated control over driving force of wheels under various unstable conditions. To sum up, the distributed electric drive automobiles reveal the automobile design concept of energy conservation, safety and environmental protection, and thus lead the tendency of future development for electric automobiles.

The change in drive form, rapid, accurate and independent control of control execution unit as well as multi-information fusion will improve immensely the dynamic control of distributed electric vehicles. The great advantages of distributed electric vehicles over traditional internal combustion engine automobiles and electric automobiles in state estimation and coordinated control are becoming a hot point in research. This dissertation will conduct intensive study on the vehicle state estimation and coordinated control.

(1) Vehicle State Estimation

Vehicle state estimation is the basis of vehicle dynamic control. Since the distributed drive motor is both the control execution unit of rapid response (which can realize rapid and accurate adjustment of driving force and braking force) and the information unit of vehicles (which can make real-time feedback of current driving moment and rotate velocity of driving wheel), the precision of vehicle state estimation will be further enhanced with the application of such feedback by information unit. In this sense, the distributed electric vehicles bring a new mode for vehicle parameter estimation, instead of the basic mode only based on inertial sensor (INS) and reference wheel velocity under traditional vehicle dynamic control system; with the combination of Global Positioning System (GPS), a series of new theories on and methods for vehicle key parameter estimation will come into being. As a result, the effect of vehicle dynamic control will be greatly improved and enhanced with the application of new state estimation system in the dynamic control over distributed electric vehicles.

(2) Coordinated Control

Distributed electric vehicles are equipped with multiple power units, with each one being independent and controllable and of rapid and accurate response. Real-time coordinated control over driving force can be realized based on the current conditions of vehicle and road surface, ensuring vehicle safety and dynamic property. In case of drive track slip, or even partial or complete failure of one or more wheels, the redundancy power source configuration system will still guarantee effective running and safe and stable driving of vehicles, satisfying demand of drivers. The coordinated control over power source of redundancy configuration, together with the real-time realization of longitudinal and lateral controlling targets of full vehicle, will improve the longitudinal driving performance and lateral stabilizing performance of vehicles, thus being of major significance to comprehensive promotion of dynamic performances of distributed electric vehicles.

1.2 Research on Status of Vehicle State Estimation

Effective vehicle dynamic control is realized through corresponding control command from full vehicle controller based on the analysis on current vehicle state according to the valid information from vehicle state estimation system, which performs online estimation of key vehicle state parameters. Real-time observation of vehicle state is the foundation of vehicle control [1], and the accuracy of such observation exerts direct impact on the control effect and property of vehicle dynamic control system. In terms of installation cost and difficulty, modern automobiles are commonly equipped with simple sensors, generally including longitudinal acceleration sensor, lateral acceleration sensor, yaw rate sensor and wheel velocity sensor. Therefore, study in this dissertation is only limited to full vehicle state variable observation with simple sensors.

It should be noted, however, that the introduction of direct signal feedback on real-time acquisition of wheel moment and rotate velocity of driving wheels by distributed electric vehicles greatly facilitates state parameter estimation, as it is a breakthrough from the basic mode only based on inertial sensor and reference wheel velocity under traditional vehicle State Estimation system. Moreover, with the development of sensor technology, sensors that were expensive and difficult in installation and debugging, such as cameras and GPS, have gradually been adopted by automobiles in large quantity, which will further facilitate the accurate observation of full vehicle state [2–5].

Currently, the vehicle state estimation mainly covers vehicle motion state (including longitudinal vehicle velocity, side slip angle, yaw rate, body pitch angle, body roll angle, etc.), tire acting force (including tire longitudinal force, tire lateral force and tire vertical force), intrinsic vehicle parameters (including mass of full vehicle, yaw inertia, location of center of mass, etc.) and road surface parameters (including road surface gradient, road surface adhesion coefficient, etc.).

Given that the research group has performed systematic observation of road surface adhesion coefficient in earlier stage [6], this dissertation will not cover specific study thereon; meanwhile, it can be recognized that the observation of longitudinal driving force is also not necessary in consideration of the known wheel driving moment of distributed electric vehicles. As a result, key variables exerting major impact on full vehicle dynamic control, including longitudinal vehicle velocity, side slip angle, yaw rate, tire lateral force, tire vertical force, mass of full vehicle and road surface gradient, are selected to be observed herein.

1.2.1 Estimation of Longitudinal Vehicle Velocity

Currently, the methods for longitudinal vehicle velocity estimation mainly include methods based on kinematics and on dynamics, wherein the former mainly covers the information on wheel velocity and INS, while the latter mainly considers information on tires, ground acting force and tire steering angle.

1.2.1.1 Longitudinal Vehicle Velocity Estimation Based on Kinematics

According to this method, longitudinal vehicle velocity is mainly determined by consideration of kinematic relation among wheel velocity, acceleration and vehicle longitudinal velocity.

(1) Longitudinal vehicle velocity estimation only based on wheel velocity
This method, which is the earliest, needs the least information from sensors. It mainly applies the fact that in-wheel velocity $v_{i,\omega}$ is approximate to vehicle velocity v_x. The Equation for in-wheel velocity $v_{i,\omega}$ is as shown in (1.1).

$$v_{i,\omega} = \omega_i R_i \qquad (1.1)$$

wherein, R_i refers to tire rolling radius while ω_i refers to tire rotate velocity. The estimated vehicle velocity can obtained only by treatment of wheel velocity signal, which mainly includes maximum wheel velocity method under braking working condition, slope method, and minimum wheel velocity method under driving working condition.

The maximum wheel velocity method, applicable to vehicle velocity estimation under braking working condition, involves real-time acquisition of four-wheel velocity and selection of maximum wheel velocity as current reference longitudinal vehicle velocity v_x [7, 8], i.e.,

$$v_x = \max\{v_{1,\omega}, v_{2,\omega}, v_{3,\omega}, v_{4,\omega}\} \qquad (1.2)$$

The estimation precision of maximum wheel velocity method depends on the precision of wheel velocity sensor and current vehicle motion state. This method is vulnerable to the wheel slip rate, which is relatively high under forced braking working condition; once overall wheel lock occurs, this method will become completely ineffective. The minimum wheel velocity method, applicable to vehicle velocity estimation under driving working condition, involves real-time acquisition of four-wheel velocity and selection of minimum wheel velocity as current reference longitudinal vehicle velocity v_x [7, 8], i.e.,

$$v_x = \min\{v_{1,\omega}, v_{2,\omega}, v_{3,\omega}, v_{4,\omega}\} \tag{1.3}$$

The estimation precision of minimum wheel velocity method depends on the precision of wheel velocity sensor and current vehicle motion state. This method is vulnerable to the wheel trackslip rate, which is relatively high under low adhesion road-surface driving working condition; once over-trackslip occurs to all wheels, this method will become completely ineffective.

The slope method is generally adopted in ABS control process. The first step of this method is to determine vehicle longitudinal initial velocity v_{x0}; and then estimate longitudinal deceleration a_x according to the vehicle working condition and road surface condition, which is the slope of vehicle velocity change [8, 9]; finally, calculate longitudinal vehicle velocity in real time according to (1.4).

$$v_x = v_{x0} + a_x t \tag{1.4}$$

The slope method requires a lot of experiments and analysis on vehicles, and deceleration value is determined based on experience and in consideration of various road surface conditions. The main disadvantage of the slope method lies in its low adaptability, and the accuracy of this method mainly depends on precision of initial velocity v_{x0} and deceleration v_{x0}. The Ref. [9] puts forward a slope method based on non-linear filtering, which is of improved adaptability but requires real-time capture of vehicle velocity peak point, thus being restrained by the error caused by selection of initial velocity.

(2) Longitudinal vehicle velocity estimation based on wheel velocity and acceleration

Currently, it is quite common to integrate the wheel velocity with acceleration information. The vehicle velocity estimation by wheel velocity requires no integral computation and longitudinal vehicle velocity can be obtained by directly multiplying wheel velocity by rolling radius; the disadvantage thereof is the large steady-state error, mainly including error in tire rolling radius, sensors, slip or trackslip. The vehicle longitudinal velocity signal can also be obtained by longitudinal integrated acceleration, which is not restrained by driving or braking working conditions. However, due to noise in acceleration signal, serious distortion may occur through long-term signal integration; in addition, initial vehicle velocity is necessary to be determined for estimation by integrated accelerated. Currently, the vehicle velocity estimation in most literatures involves the judgment for accuracy of wheel velocity signal and acceleration signal according to the current vehicle state and integrated observation of the two types of signals, which can be expressed as shown in (1.5) [7].

$$v_x(k) = \frac{\sum_{i=1}^{4} k_i v_{i,\omega} + k_{a_x}(v_x(k-1) + \Delta T a_x)}{\sum_{i=1}^{4} k_i + k_{a_x}} \tag{1.5}$$

wherein, k_i refers to the wheel velocity weight coefficient, k_{a_x} refers to the acceleration weight coefficient, ΔT refers to the time step, and the longitudinal vehicle velocity equals to the weighted average of integration of wheel velocity and acceleration.

A extensively used mathematic expression in Eq. (1.5) is Kalman Filtering (KF) [10, 11]. Its recurrence equation is created by virtue of the relation between acceleration and velocity, as shown in Eq. (1.6).

$$v_x(k) = v_x(k-1) + a_x \Delta T + q_{a_x} \tag{1.6}$$

The KF measurement equation can be created with wheel velocity information, as shown in Eq. (1.7).

$$v_\omega(k) = v_x(k) + r_\omega \tag{1.7}$$

wherein, q_{a_x} and r_ω refer to noise in acceleration information and wheel velocity information respectively.

The common methods focus on the determination of noise amplitude according to the vehicle state: the louder the noise is, the lower the reliability of state equation will be, i.e., the smaller the weight of corresponding item in Eq. (1.5) will be.

The Ref. [12] put forward a noise regulation method based on reliable index sensor fusion (RISF), according to which the noise in state equation and observation equation can be expressed by Eq. (1.8).

$$q_\omega \quad or \quad r_{a_x} = \sum_i c_i \exp(d_i x_i) \tag{1.8}$$

wherein, x_i refers to sensor information, while c_i and d_i are constants determined through experience and experiments.

The Ref. [13] considered the estimation errors in noise regulation caused by tire rolling radius error, sudden change in wheel velocity, wheel trackslip or slip, etc., and sets the noise regulation method based on rules; the [14] puts forward the method for weighting function regulation by fuzzy logic based on the Ref. [13]—which assumed that the bigger the difference value $|v_\omega - v_x|$ between in-wheel velocity and estimated vehicle velocity or the difference value $|a_\omega - a_x|$ between in-wheel acceleration and vehicle acceleration is, the more serious the tire trackslip or slip will be, i.e., the smaller the weight of wheel velocity information will be, or otherwise, the bigger weight of the same will be; the Ref. [15] developed the above method and applies it to the vehicle steering process after considering the influence factors of steering angle of front wheel on noise, thus obtaining higher observation precision.

Besides the above-mentioned typical observation methods, some scholars also put forward other estimation methods by virtue of non-linear observer [16], sliding-mode observer [17] and Luenberger observer [18], which, through considering the impact of vehicle non-linear property on observation precision, are of higher observation precision while of ordinary observation timeliness; meanwhile, system observability also needs to be ensured.

1.2.1.2 Estimation of Longitudinal Vehicle Velocity Based on Dynamics

This method involves the estimated vehicle velocity from the longitudinal vehicle velocity output obtained by establishment of reasonable vehicle longitudinal dynamic model, with the input being the information on driving force and braking force acquired. The estimation precision of this method depends on the precision of vehicle driving force and braking force, and is greatly affected by the vehicle longitudinal dynamic model and tire model [19]. Therefore, the precision will be enhanced with reasonable vehicle longitudinal dynamic model and tire model as well as obtaining of accurate longitudinal driving force and braking force. In selection of vehicle model, the impacts on both accuracy and timeliness shall be considered; the vehicle model with longitudinal, lateral, roll and axle load transfer being taken into consideration is the most common, as shown typically in the Ref. [20].

1.2.2 Estimation of Side Slip Angle

As an important index for vehicle stability, side slip angle has been a hot point in estimation. Unlike longitudinal vehicle velocity, lateral vehicle velocity is generally lower; however, small error in lateral velocity will lead to big error in side slip angle observation; meanwhile, useful information will easily be annihilated in the background noise due to the vulnerability of lateral velocity to vehicle yaw and longitudinal motion. Similar to the classification of observation methods for vehicle longitudinal velocity, the observation methods for side slip angle can be based on kinematics and dynamics. In view of the fact that estimation of side slip angle lies mainly in estimation of lateral vehicle velocity [20] this dissertation will integrate the estimation of lateral velocity with that of side slip angle in literature summary.

1.2.2.1 Estimation of Side Slip Angle Based on Kinematics

With the observer designed based on kinematics, the side slip angle estimation can be directly conducted through the kinematic relation between side slip angle and yaw rate, lateral acceleration, wheel velocity and vehicle velocity.

(1) Estimation of side slip angle by side slip angle velocity integration

The estimation of side slip angle β with integration of side slip angle velocity $\dot{\beta}$ is the most common method based on kinematics. The Ref. [21] put forward the method for calculating side slip angle velocity $\dot{\beta}$ by (1.9).

$$\dot{\beta} = \frac{1}{1 + \beta^2} \left(\frac{\dot{v}_y}{v_x} - \gamma - \beta \frac{\dot{v}_x}{v_x} - \beta^2 \gamma \right) \qquad (1.9)$$

wherein, γ refers to the vehicle yaw rate and v_y refers to the lateral velocity. Generally, Eq. (1.9) can be applied only when pitch angle and roll angle are very small, and is applicable only to the steady-state driving working condition with relatively small longitudinal acceleration and lateral acceleration.

In case of small side slip angle and small change in vehicle longitudinal velocity, Eq. (1.9) can be simplified [12, 21, 22] and the estimation can be performed through Eq. (1.10).

$$\beta(t) = \beta_0 + \int_0^t \left(\frac{\dot{v}_y}{v_x} - \gamma \right) dt \qquad (1.10)$$

With Eq. (1.10)—the robustness will be better and the change in model parameter (mass, tire cornering stiffness, etc.) exerts almost no impact on the estimation effect; however, this method heavily relies on INS accuracy and initial value selection, and the integration accumulation error will exert serious effect on the estimation under long-term working condition, so the sensor signal must be corrected. This method is only applicable to the approximately constant longitudinal vehicle velocity and relatively small side slip angle.

(2) Estimation of side slip angle with rotate velocity and steering angle of front wheel

The Ref. [23] shows a solution to estimation of vehicle lateral velocity by virtue of the kinematic relation among steering angle of front wheel, wheel velocity and vehicle lateral velocity, as shown in Eq. (1.11).

$$v_y = -l_f \gamma + \frac{v_{1,\omega} + v_{2,\omega}}{2} \sin(\delta - \alpha_f) \qquad (1.11)$$

wherein, l_f refers to the distance from front axle and center of mass, α_f refers to the slip angle of front axle, and δ refers to the steering angle of front wheel. An advantage of this method lies in the cancellation of integration link, thus avoiding accumulative error; however, this method is restrained by wheel velocity to a large extent, which is only applicable to small tire trackslip and sideslip in general.

(3) Estimation of side slip angle by integration of longitudinal and lateral acceleration Kalman filtering observation with two-degree of freedom vehicle kinematic model is a common observation method. Many Refs. [4, 12, 16, 24] created state recurrence equation by virtue of in-vehicle INS, as shown in Eq. (1.12).

$$\begin{bmatrix} \dot{v}_x \\ \dot{v}_y \end{bmatrix} = \begin{bmatrix} 0 & \gamma \\ -\gamma & 0 \end{bmatrix} \begin{bmatrix} v_x \\ v_y \end{bmatrix} + \begin{bmatrix} a_x \\ a_y \end{bmatrix} + \begin{bmatrix} q_{a_x,\gamma} \\ q_{a_y,\gamma} \end{bmatrix} \tag{1.12}$$

Even with the state recurrence equation, Eq. (1.12) does not contain measurement equation. There are many ways of creating measurement equation, such as establishing relations among wheel velocity, vehicle velocity and yaw rate [23], obtaining lateral velocity with two-degree of freedom vehicle model [12] and obtaining longitudinal and lateral velocity through GPS [4]. It should be noted that this method, relying on the integration of longitudinal and lateral acceleration sensors, requires the consideration of sensor static deviation compensation and noise property, and the necessity of self-adaptation design for noise $q_{a_x,\gamma}$ and $q_{a_y,\gamma}$.

1.2.2.2 Estimation of Side Slip Angle Based on Dynamics

This method involves the estimation from the state information or output information of vehicle model built, and by introducing the information on tire acting force and vehicle acting force to the observation system firstly. This method requires high model precision, rather than that of in-vehicle INS. The following modes are the most common for this method.

(1) Estimation of side slip angle with single-track vehicle model
Single-track vehicle model is the most common state observation model, which is adopted in many Refs. [12, 18, 22–29]. The model can be expressed by the state recurrence equation shown in Eq. (1.13).

$$\dot{x} = Ax + Bu \tag{1.13}$$

$$x = \begin{bmatrix} \beta \\ \gamma \end{bmatrix}, A = \begin{bmatrix} a_{11} & a_{12} \\ a_{21} & a_{22} \end{bmatrix} = \begin{bmatrix} \dfrac{-(C_f + C_r)}{mv_x} & \dfrac{-l_fC_f + l_rC_r}{mv_x^2} - 1 \\ \dfrac{-l_fC_f + l_rC_r}{I_z} & \dfrac{-l_f^2C_f - l_r^2C_r}{I_zv_x} \end{bmatrix},$$

$$B = \begin{bmatrix} b_1 \\ b_2 \end{bmatrix} = \begin{bmatrix} \dfrac{C_f}{mv_x} \\ \dfrac{l_fC_f}{I_z} \end{bmatrix}, u = \delta$$

wherein, C_f refers to the cornering stiffness of front axle; C_r refers to the cornering stiffness of rear axle; l_r refers to the distance from rear axle to center of mass; I_z refers to the yaw rotational inertia of vehicle.

Under steady-state steering working condition, the side slip angle of vehicle is the linear function for steering angle of front wheel, based on which the Refs. [27, 28] created the function for side slip angle to steering angle δ of front wheel, as shown in Eq. (1.14).

$$\beta = \frac{1 - m l_f v_x^2 / (2 l l_r C_r)}{1 - m v_x^2 (l_f C_f - l_r C_r) / (2 l^2 C_f C_r)} \cdot \frac{l_r}{l} \delta \tag{1.14}$$

wherein, l refers to the axle base. Since Eq. (1.14) is only applicable to estimation of side slip angle during steady-state steering, the estimated value will not be accurate in case of visible sideslip or vehicle acceleration.

The Ref. [30] provided the Equation for calculating side slip angle with lateral acceleration and yaw rate under steady-state working condition, i.e., Eq. (1.15).

$$\beta = \frac{l_r}{v_x} \gamma - \frac{m l_f}{2 l C_r} a_y \tag{1.15}$$

The single-track vehicle model focuses on the impact of tire yaw force on vehicle lateral motion, instead of the tire longitudinal force. As the tire cornering stiffness is subject to various impact factors and the non-linear effect becomes significant in case of strong yaw force, self-adaptation of tire cornering stiffness is required to improve estimation effect [25].

(2) Estimation of side slip angle with double-track vehicle model
The double-track vehicle model involving longitudinal, lateral and yaw motions, which was adopted in the Refs. [7, 16, 20, 31–34]—has been widely applied in the observation of side slip angle. This method can be expressed by the state recurrence equation shown in Eq. (1.16).

$$\begin{cases} \dot{v}_x = v_y \gamma + \frac{1}{m}((F_{1,x} + F_{2,x})\cos\delta) - (F_{1,y} + F_{2,y})\sin\delta + (F_{3,x} + F_{4,x})) \\ \dot{v}_y = -v_x \gamma + \frac{1}{m}((F_{1,x} + F_{2,x})\sin\delta) + (F_{1,y} + F_{2,y})\cos\delta + (F_{3,y} + F_{4,y})) \\ \dot{\gamma} = \frac{1}{I_z}[(F_{1,x} + F_{2,x})l_f \sin\delta + (F_{1,y} + F_{2,y})l_f \cos\delta - (F_{3,y} + F_{4,y})l_r \\ \quad + \frac{1}{2}(F_{2,x} - F_{1,x})b_f \cos\delta - \frac{1}{2}(F_{2,y} - F_{1,y})b_f \sin\delta + \frac{1}{2}(F_{4,x} - F_{3,x})b_r \cos\delta] \end{cases} \tag{1.16}$$

wherein, b_f refers to the front wheel tread, while b_r refers to the rear wheel tread. Compared with single-track vehicle model, double-track vehicle model, involving the consideration of impact of longitudinal driving force on longitudinal velocity and yaw rate, requires higher model precision, and especially relies on the accuracy of tire yaw force. Some scholars have adopted more advanced full vehicle dynamic models for more accurate estimation of side slip angle; however, the timeliness and stability of such models in practical application are still pending for verification.

1.2.3 Yaw Rate Estimation

The yaw rate can directly be obtained from yaw rate sensors owing to universal application of such sensors. In addition, yaw rate observation can be performed by virtue of the kinematic relation between wheel velocity and yaw rate [35, 36]. The Ref. [26] provided the method for yaw rate estimation with rear wheel velocity, as shown in Eq. (1.17).

$$\gamma = \frac{v_{4,\omega} - v_{3,\omega}}{b_r} \tag{1.17}$$

The Ref. [23] extended the calculation by introducing the steering angle of front wheel, involving the estimation with four-wheel velocity. This method can be effective to a certain extent in case of curve driving, as shown in Eq. (1.18).

$$\gamma = \frac{v_{4,\omega} - v_{3,\omega}}{2b_r} + \frac{v_{2,\omega} - v_{1,\omega}}{2b_f} \cdot \frac{\cos(\delta - \alpha_f)}{\cos \alpha_f} \tag{1.18}$$

In addition, the estimation can also be conducted according to the state recurrence equation shown in Eq. (1.13) through observation of lateral acceleration with a Kalman filter [37].

In general, yaw rate is the easiest to be estimated among vehicle motion state parameters; good observation effect can be obtained as long as with the fusion of wheel velocity with information from yaw rate sensor. Currently, there is a trend of combined observation in terms of vehicle dynamic state parameters. It will be most significant in the future that combined observation of longitudinal vehicle velocity, side slip angle and yaw rate is performed through uniform observer built with vehicle models.

1.2.4 Estimation of Tire Lateral Force

The estimation of tire lateral force relies on tire models. Currently common tire model building mechanism include two types: one is the theoretical model represented by Brush model, which obtains the relation among relevant variables through analysis on tire structure and tire deformation; the other is the experience or semi-experience model represented by Dugoff model and Magic Formula Model. It should be noted that the experience model is built through a lot of experiments and data statistics analysis, while semi-experience model involves analysis result of experimental data besides theoretical analysis on tire property.

(1) Brush model

Brush model, a typical physical model [38] put forward by Dugoff et al., is a simplified theoretical model assumed on the basis of "elastic tire surface and stiff tire body", focusing on the elasticity of tires on the set tire surface with "brush" deformation and regarding the tire body as rigid body [39]. Pacejka further developed the Brush model [40] and showed it as Eq. (1.19).

$$F_y = \begin{cases} -3\mu F_z\rho_y(1 - |\rho_y| + \dfrac{1}{3}\rho_y^2), & |\alpha| \le |\alpha_{sl}| \\ -\mu F_z\mathrm{sgn}(\alpha), & |\alpha| > |\alpha_{sl}| \end{cases} \tag{1.19}$$

wherein, $\rho_y = \theta_y\sigma_y\alpha_{sl} = \arctan(1/\theta_y)\theta_y = C_\alpha/(3\mu F_z)\sigma_y = \tan\alpha$, and μ refers to the road surface adhesion coefficient.

(2) Dugoff model

Dugoff model is the semi-experience model [41] put forward by Dugoff et.al. in 1970 based on theoretical derivation and experimental data, describing the relation among tire slip rate, slip angle, longitudinal force and lateral force, reflecting well the non-linear properties of tires and applying to longitudinal and lateral joint slip/trackslip working conditions. Currently, Dugoff model has been successfully applied to the ESP system of Bosch Corporation [21]. It can be expressed by Eq. (1.20).

$$F_y = C_\alpha \frac{\tan\alpha}{1 + \sigma_x} f(\lambda) \tag{1.20}$$

$$\lambda = \frac{\mu F_z(1 + \sigma_x)}{2((C_\sigma\sigma_x)^2 + (C_\alpha\tan\alpha)^2)^{1/2}}, \quad f(\lambda) = \begin{cases} (2-\lambda)\lambda, & \lambda < 1 \\ 1, & \lambda \ge 1 \end{cases}$$

wherein, C_σ refers to the longitudinal slip stiffness of tire; C_α refers to the cornering stiffness of tire; σ_x refers to the longitudinal slip/trackslip rate and α refers to the slip angle of tire.

(3) Magic Formula Model

Magic Formula Model was built by Pacejka et.al. in the 1990 s through a lots of tire experiments [40], which completely expresses the mechanical properties of longitudinal force, lateral force and aligning moment of tire with a series of Equations, thus being recognized as the most accurate tire model currently and widely applied in vehicle model building. It can be expressed by Eq. (1.21).

$$\begin{cases} y = D \sin[C \arctan\{Bx - E(Bx - \arctan(Bx))\}] \\ Y = y + S_V \\ x = X + S_H \end{cases} \quad (1.21)$$

wherein, Y refers to the longitudinal force, lateral force or aligning moment; X refers to slip rate or slip angle; S_V refers to longitudinal axis offset; S_H refers to horizontal axis offset and B, C, D and E are subject to the impact of tire vertical force, tire camber angle, etc. Nagai et al. simplified the Magic Formula Model and put forward the tire lateral force model with change in longitudinal driving force, vertical load transfer and road surface adhesion coefficient being taken into consideration. This model, with less fitting parameters, can better approximate the Magic Formula Model in case of insignificant non-linear effect [42, 43], as shown in Eq. (1.22).

$$\begin{cases} F_y = \dfrac{2}{\pi} K_x \mu F_z \arctan(\dfrac{\pi C_\alpha \alpha}{2\mu F_z}) \\ K_x = \sqrt{1 - \dfrac{F_x^2}{\mu^2 F_z^2}} \end{cases} \quad (1.22)$$

German scholar Kiencke put forward a tire model with the impact of vertical force is taken into consideration, which is also regarded as a simplified form of Magic Formula Model, as shown in Eq. (1.23).

$$F_y = k_{red}(1 - \dfrac{F_z}{\xi_1})F_z \arctan(\xi_2 \alpha) \quad (1.23)$$

wherein, ξ_1, ξ_2 and k_{red} are all model parameters.

In general, theoretical models are of less parameters and clarified physical significance but lower fitting precision, while experience models are of complex forms and more parameters but higher fitting precision. With the development of modern computing technology, the computing speed becomes faster and faster and the complex experience model is seeing more and more applications [44, 45]. Currently, the estimation of tire lateral force relies more and more on accurate tire models. The unique independent electric drive structure of distributed electric vehicles is quite favorable for more accurate estimation of tire lateral force.

1.2.5 Estimation of Tire Vertical Force

Vertical force is very significant to vehicle dynamic control, such as lateral force estimation [43], suspension control [46] and anti-rollover control [47–49].

The simplest method for vertical force observation is the axle load transfer calculation with longitudinal and lateral acceleration [6, 28, 43], as shown in Eq. (1.24).

$$
\begin{bmatrix} F_{1,z} \\ F_{2,z} \\ F_{3,z} \\ F_{4,z} \end{bmatrix} = \begin{bmatrix} l_r \\ l_r \\ l_f \\ l_f \end{bmatrix} \frac{mg}{2l} + \begin{bmatrix} -1 \\ -1 \\ 1 \\ 1 \end{bmatrix} \frac{ma_x h_g}{2l} + \begin{bmatrix} -l_r \\ l_r \\ -l_f \\ l_f \end{bmatrix} \frac{ma_y h_g}{lb}
\tag{1.24}
$$

wherein, h_g refers to the height of vehicle center of mass, l refers to the axle base and b refers to wheel tread. This method features simple calculation and strong timeliness and is applicable to scenes undemanding for vertical force; however, since this method neglects the impact of axle load transfer and road surface gradient caused by vehicle body rolling, the estimation effect is limited.

The precision of vertical force observation can be enhanced by introducing rolling motion of vehicle body. A common method is observing the roll angle of vehicle body before observing vertical force, by which higher precision can be obtained under the condition of strong lateral excitation for vehicles. The Refs. [50–52] adopted the rolling motion equation shown in Eq. (1.25) for observation of roll angle of vehicle body and roll rate, during which lateral velocity, yaw rate and lateral acceleration are employed.

$$
\begin{cases} I_z \ddot{\phi} + C_{\text{roll}} \dot{\phi} + K_{\text{roll}} \phi = mh_{\text{roll}} a_{y,m} \\ a_{y,m} = \dot{v}_y + \gamma v_x + g\phi + h_r \ddot{\phi} \end{cases}
\tag{1.25}
$$

wherein, h_r refers to the height of vehicle center of mass relative against roll center; $a_{y,m}$ refers to the measured value of lateral acceleration sensor; K_{roll} refers to the roll angle stiffness and C_{roll} refers to the roll damping. The Ref. [53] created the roll angle observation equation by virtue of the relation between suspension displacement sensor and roll angle, as shown in Eq. (1.26).

$$
\phi = \frac{z_2 - z_1}{2b_f} + \frac{z_4 - z_3}{2b_r} - \frac{ma_{y,m} h_g}{k_{y,t}}
\tag{1.26}
$$

wherein, $k_{y,t}$ refers to the roll stiffness of tire system.

Assume that the axle load transfer caused by vehicle body rolling is $F_{z,\phi}$, then the estimation of vertical force in case of vehicle body rolling is as shown in Eq. (1.27).

$$
\begin{bmatrix} F_{1,z} \\ F_{2,z} \\ F_{3,z} \\ F_{4,z} \end{bmatrix} = \begin{bmatrix} l_r \\ l_r \\ l_f \\ l_f \end{bmatrix} \frac{mg}{2l} + \begin{bmatrix} -1 \\ -1 \\ 1 \\ 1 \end{bmatrix} \frac{ma_x h_g}{2l} + \begin{bmatrix} -l_r \\ l_r \\ -l_f \\ l_f \end{bmatrix} \frac{ma_y h_g}{lb} + \begin{bmatrix} -1 \\ 1 \\ -1 \\ 1 \end{bmatrix} \frac{F_{z,\phi}}{2}
\tag{1.27}
$$

With the development of vehicle manufacturing technology, there is more and more applicable information to full vehicle controller and the application of various types of sensors in the combined vertical force observation will be one of the future trends.

1.2.6 Estimation of Mass and Gradient

In the process of vehicle running, mass of full vehicle and road surface gradient are significant parameters to vehicle driving and braking. Accurate mass of full vehicle and road surface gradient are always necessary for active safety control of vehicles [54]. In case of large deviation between estimated mass of full vehicle and the actual value, the model on which vehicle controller is relied will become inaccurate and the effect of various types of active safety control will become poor. As for gradient, its impact shows the greatest significance in heavy commercial vehicles and off-road vehicles. Even slight gradient will cause great change in desired driving/braking power. In this sense, it is very necessary to conduct estimation of mass (short term for mass of full vehicle) and gradient. Both gradient and mass are important parameters for vehicle longitudinal dynamic control and the observation methods for them are closely related, which are mainly divided into kinematic methods and dynamic methods.

1.2.6.1 Estimation of Mass and Gradient Based on Kinematics

Kinematics is mostly applied to gradient estimation. Currently, many literatures have made gradient estimation with in-vehicle GPS and barometric pressure sensor. The Refs. [54, 55] employed GPS in current gradient observation; in case of two in-vehicle GPS sensors, relative location information of the two sensors, which covers information on road surface gradient and vehicle pitching motion, can be determined by tracing the carrier phase of each sensor. Compared with the vehicle pitching vibration, the change in road surface gradient is so slow that the information frequency thereof is generally lower than 0.5 Hz, and such information can be obtained by low-pass filtering; in case of only one GPS sensor, the estimation of instantaneous road surface gradient can be performed by comparing the information on vertical velocity and horizontal velocity from GPS, and relatively reliable road surface gradient can also be obtained by low-pass filtering. The Ref. [56] involved estimation of road surface gradient by Kalman filter with known mass, based on the measurement of altitude with GPS sensor and barometric pressure sensor; the Refs. [57, 58] extended the above method and introduce the dynamic relation between driving force and longitudinal acceleration in estimation of road surface gradient; the Refs. [56, 57] also considered the impact of weather while noting the variance of barometric pressure with the variance of altitude; the Ref. [59] provided further mathematic relation between barometric pressure and altitude. An advantage of this method is full utilization of existing sensors for vehicles without the need of any other sensors. However, the disadvantages thereof are also quite noticeable: on one hand, the low GPS frequency is a major problem for estimation of road surface gradient by GPS and small GPS velocity error may lead to big estimation error; on the other hand, the barometric pressure sensor features low reliability and signal-to-noise ratio, for it is vulnerable to various external factors, including temperature, wind speed and weather, and the barometric pressure varies very slowly with the variance of altitude. The Refs. [60, 61] put forward a simple method for observing road surface gradient, which takes advantage of the kinematic relation among longitudinal acceleration sensor signal $a_{x,m}$, driving acceleration $\dot{v}_x$ and road surface gradient θ—as shown in Eq. (1.28).

$$\theta = \frac{a_{x,m} - \dot{v}_x}{g} \tag{1.28}$$

The estimation method shown in Eq. (1.28) features simple calculation and good timeliness, but requires high sensor precision, i.e., accurate information from vehicle driving acceleration sensor and longitudinal acceleration sensor, wherein the static deviation of the latter exerts great impact on estimation effect. With the progress in electronic map technology and GPS technology, it will be a significant trend for estimation of road surface gradient through real-time combination of GPS and local electric map [62].

1.2.6.2 Estimation of mass and gradient based on dynamics

Dynamics is mostly applied in combined estimation of mass and gradient. Vahidi et al. put forward a method for simultaneous estimation of mass and gradient [63, 64] by least square method [65–67]. A longitudinal dynamic equation is employed, as shown in Eq. (1.29).

$$\dot{v}_x = \frac{1}{m}(F_x - F_w) - g(f \cos\theta + \sin\theta) \tag{1.29}$$

wherein, F_x refers to the longitudinal driving force of full vehicle and F_w refers to wind drag. Make $\varphi_1 = 1/m$, $\varphi_2 = g(f \cos\theta + \sin\theta)$: since the mass is always constant during vehicle motion at a time and φ_1 is constant value, estimation by least square method is feasible; since the gradient is time-varying and φ_2 is also time-varying, the estimation by least square method φ_2 requires the introduction of forgetting factors.

The Ref. [68] developed the above method by applying self-adaptation observer [69] in simultaneous observation of mass and gradient and designing self-adaptation law by Lyapunov method, thus ensuring the stability of observation. By way of this, certain estimation effect can be achieved in case of violent fluctuation such as step change in road surface gradient, by correcting the estimated value in real time according to the self-adaptation law. The Refs. [70–72] utilized Kalman filter in the combined estimation of mass and gradient, as shown in Eq. (1.29). Though the observability of this kind of observer has been proved, the estimation effect in case of great change in gradient is still pending for further investigation.

The Refs. [54, 55, 73] created the observation equation by longitudinal dynamic method with known gradient, and estimate the mass of full vehicle through Kalman filtering, with the precision depending on estimated precision of road surface gradient-even small gradient error may cause big error in mass estimation.

Currently, almost all dynamic methods require a command of various factors, such as drag coefficient, rolling resistance coefficient and current wind speed, and have not achieved good coupling in estimation of mass and gradient.

1.3 Research Status of Coordinated Control

Distributed electric vehicles are equipped with multiple independent and controllable power units, which can perform coordinated control over driving force according to current conditions of vehicles and road surfaces, thus ensuring the dynamic property and safety of vehicles. Coordinated control is important particularly when drive failure or trackslip of vehicles occurs, so distributed electric vehicles can satisfy drivers' demand through coordinated control over redundancy power source, thus ensuring effective, safe and stable driving, while since traditional electric vehicles are only equipped with single power source, coordinated control over multiple drive units can not be realized in the same case; instead, vehicles have to be subject to reduction of driving motor moment, or even motor shutdown and immediate vehicle stop. The study on coordinated control of distributed electric vehicles focuses mainly on several perspectives, namely, lateral control target, driving force control allocation and motor differential compensation.

1.3.1 Lateral Control Target

The basic principle of vehicle lateral control lies in real-time monitoring of vehicle motion state parameters, reasonable target and execution of lateral control by full vehicle controller and further regulation of lateral motion state of vehicles. Traditional vehicles perform this function only through cooperation of braking system and motor, which may cause slow vehicle response speed and inaccurate control, while distributed electric vehicles can realize real-time tracing of expected moment with independent and controllable driving force moment of each wheel. Therefore, lateral stability property of vehicles can be greatly improved by combing the characteristics of distributed electric drive and reasonable lateral control target.

1.3.1.1 Analysis on Lateral Control Target Based on Two-Degree of Freedom Vehicle Model

In order to describe vehicle working conditions in case of strong lateral excitation [22, 25, 30, 42, 43], two-degree of freedom vehicle model is commonly used for designing lateral stability control target in terms of vehicle dynamic control. The vehicle dynamic control equation with direct yaw control (DYC) being taken into consideration is as shown in Eq. (1.30)—which increases the yaw control item EM compared with Eq. (1.13). Through direct control over yaw rate by yaw moment, indirect control over side slip angle can be realized.

$$\dot{x} = Ax + Bu + EM \tag{1.30}$$

wherein,

$$E = [0, e_2]^{\mathrm{T}}, M = M_z$$

M_z refers to the direct yaw moment. Through Laplace transformation, the transfer function for vehicle side slip angle and yaw rate subject to excitation of steering angle of front wheel can be obtained, as shown in Eqs. (1.31) and (1.32).

$$\beta(s) = \frac{(b_1 s - b_1 a_{22} + b_2 a_{12})\delta(s) + e_2 a_{12} M_z(s)}{s^2 - (a_{11} + a_{22})s + a_{11}a_{22} - a_{21}a_{12}} \tag{1.31}$$

$$\gamma(s) = \frac{(b_2 s + b_1 a_{21} - b_2 a_{11})\delta(s) + e_2(s - a_{11})M_z(s)}{s^2 - (a_{11} + a_{22})s + a_{11}a_{22} - a_{21}a_{12}} \tag{1.32}$$

For vehicles without DYC, the direct yaw moment is 0, therefore the transfer function from steering angle of front wheel to side slip angle and yaw rate will be degraded to Eqs. (1.33) and (1.34).

$$\beta(s) = \frac{(b_2 s - b_1 a_{22} + b_2 a_{12})}{s^2 - (a_{11} + a_{22})s + a_{11}a_{22} - a_{21}a_{12}}\delta(s) \tag{1.33}$$

$$\gamma(s) = \frac{(b_2 s + b_1 a_{21} - b_2 a_{11})}{s^2 - (a_{11} + a_{22})s + a_{11}a_{22} - a_{21}a_{12}}\delta(s) \tag{1.34}$$

Without DYC, the steady-state response of side slip angle and yaw rate can be expressed by Eqs. (1.35) and (1.36).

$$\beta_{0,n} = \frac{-b_1 a_{22} + b_2 a_{12}}{a_{11}a_{22} - a_{21}a_{12}}\delta \tag{1.35}$$

$$\gamma_{0,n} = \frac{b_1 a_{21} - b_2 a_{11}}{a_{11}a_{22} - a_{21}a_{12}}\delta \tag{1.36}$$

When the steady-state side slip angle is restrained to 0 by DYC, the required direct yaw moment can be calculated through Eq. (1.31)—as follows:

$$M_z = \frac{a_{22}b_1 - a_{12}b_2}{a_{12}e_2}\delta \tag{1.37}$$

Substitute Eq. (1.37) into Eqs. (1.31) and (1.32)—and then the transfer function from steering angle of front wheel to side slip angle and yaw rate can be obtained when the side slip angle is restrained to 0, as follows:

$$\beta(s) = \frac{b_1 s}{s^2 - (a_{11} + a_{22})s + a_{11}a_{22} - a_{21}a_{12}} \delta(s) \tag{1.38}$$

$$\gamma(s) = -\frac{b_1}{a_{12}} \cdot \frac{-a_{22}s + (a_{11}a_{22} - a_{12}a_{21})}{s^2 - (a_{11} + a_{22})s + (a_{11}a_{22} - a_{12}a_{21})} \delta(s) \tag{1.39}$$

It can be seen from Eqs. (1.38) and (1.39) that the steady-state response of side slip angle and yaw rate with DYC can be expressed by Eqs. (1.40) and (1.41).

$$\beta_{0,w} = 0 \tag{1.40}$$

$$\gamma_{0,w} = -\frac{b_1}{a_{12}}\delta \tag{1.41}$$

The above is the analysis on transfer function and steady state response of vehicle side slip angle and yaw rate with/without active control, wherein many variables can be used as control targets, which will be discussed hereunder.

1.3.1.2 Direct Yaw Moment Calculation Based on Yaw Rate Control

Compared with side slip angle, yaw rate is easier to obtain. Many literatures take controlling yaw rate approximate to target value as the target for yaw control, based on which the desired direct yaw moment is set. The Refs. [74, 75] adopted the direct yaw moment control with proportional gain, as shown in Eq. (1.42).

$$M_z = K_p(\gamma - \gamma_d) \tag{1.42}$$

The target yaw rate γ_d in Eq. (1.42) is designed based on the steady-state yaw rate $\gamma_{0,n}$ without active control; in order to reflect the vehicle properties more accurately, the Ref. [74] also obtained more accurate value of γ_d through experiments. This method allows effective tracing of expected yaw rate in case of steering acceleration with low adhesion coefficient of road surface. The Ref. [76] adopted fuzzy control over yaw rate, with the fuzzy controller input being yaw rate error $\gamma - \gamma_d$ and the rate of change $\dot{\gamma} - \dot{\gamma}_d$, and output being the direct yaw moment. The target yaw rate γ_d employed herein is the correction of $\gamma_{0,n}$−n, as shown in Eq. (1.43).

$$\gamma_d = \frac{v_x}{l}\delta + r_c(\delta, v_x) \tag{1.43}$$

wherein, r_c refers to the correction parameter.

According to Eq. (1.43)—it is assumed that the vehicles perform neutral steer, and neural network method is adopted for regulating the target yaw rate, with the regulating parameters being δ and v_x. This control method ensures the stability in straight driving and steering, and improves the vehicle performance in passing through poor road surface. The Ref. [77] calculated the expected direct yaw moment by sliding-mode control method, with the designed sliding-mode surface composed of yaw rate error $\gamma - \gamma_d$ and the rate of change $\dot{\gamma} - \dot{\gamma}_d$. This design method ensures the robustness and stability of control, except that the high-frequency buffeting caused by sliding mode shall be considered and eradicated. It should be noted that during yaw rate control, the designed target yaw rate is also restrained by road surface adhesion [78, 79], as shown in Eq. (1.44).

$$|\gamma_d| \leq \zeta_\gamma \frac{\mu g}{v_x} \tag{1.44}$$

wherein, the factor ζ_γ involves the impact of combined action by longitudinal and lateral acceleration and μ refers to the road surface adhesion coefficient.

1.3.1.3 Direct Yaw Moment Calculation Based on Combined Control over Yaw Rate and Side Slip Angle

With the progress in lateral stability control technology and side slip angle observation technology, the control only over yaw rate can not satisfy the demand for control any longer, while control over side slip angle has become feasible. The vehicle lateral stability will be improved to a larger extent by conducting side slip angle control while introducing the yaw rate control [80]. Currently, there have been various combined control methods for yaw rate and side slip angle.

The Refs. [81–84] selected fuzzy control method for control over yaw rate and side slip angle, with the fuzzy controller input being yaw rate error $\gamma - \gamma_d$ and side slip angle error $\beta - \beta_d$, and target yaw rate being both $\gamma_{0,n}$. However, the target side slip angle β_d selected in various methods are slightly different from one another: the Refs. [81–83] took the target side slip angle as 0, while the Ref. [84] as $\beta_{0,n}$.

The Ref. [85] conducted simultaneous control over yaw rate error $\gamma - \gamma_d$ and side slip angle error $\beta - \beta_d$ by H_∞ control method, and first order delay property is also considered for the designed target yaw rate γ_d on the basis of $\gamma_{0,n}$—as shown in Eq. (1.45).

$$\gamma_d = \frac{1}{1 + \tau_\gamma s} \gamma_{0,n} \tag{1.45}$$

wherein, τ_γ refers to time constant. The Refs. [80, 86] designed the method for simultaneous tracing of target side slip angle and yaw rate based on sliding-mode theory, with the target side slip angle selected being the second order transfer function shown in Eq. (1.33)—and target yaw rate being the same with that of Eq. (1.45).

The Ref. [25] calculate the direct yaw moment by optimum control method and traced the expected side slip angle and yaw rate; however, the setting of side slip angle and yaw rate is not mentioned therein. Similar to Eq. (1.44) for restrained yaw rate, the designed target side slip angle shall also be subjected to road surface adhesion during side slip angle control [78, 79], as shown in Eq. (1.46).

$$|\beta_d| \leq |\frac{a_y}{a_{y,\max}}(\frac{l_r g}{v_x} - \frac{v_x}{C_f})| \tag{1.46}$$

wherein (1.46)—$a_{y,\max}$ refers to the maximum lateral velocity under the current vehicle velocity and steering radius.

Generally speaking, vehicle stability performance can be improved more under simultaneous control over side slip angle and yaw rate than under separate control over yaw rate; however, the controlling effect is quite vulnerable to the estimated precision of side slip angle.

1.3.1.4   Direct Yaw Moment Calculation Based on Combined Feedforward and Feedback Control

In case of feedforward control, system running condition of the next stage can be predicted through recognition of external environment and current state, and then the controlled values can be given and potential error can be compensated. Feedforward control requires certain knowledge on the controlling object and features rapid response or even control completion before occurrence of deviation; however, complex control can not be realized only by feedforward, for there's always disturbance and deviation during control, which requires real-time correction by introducing feedback control. The combined feedforward and feedback control structure, well solved the problems of low control precision, weak timeliness and high feedback overshoot, and has been widely applied in vehicle dynamic control [87]. Nagai et al. applied the combined feedforward and feedback control structure in control over direct yaw moment [30, 42, 43, 88–93], with the control structure as shown in Fig. 1.1.

According to this solution, the target of feedforward is the desired direct yaw moment when steady-state response is satisfied, and the feedforward direct yaw moment M_{ff} generally depends on the steering angle δ of front wheel and feedforward control gain G_{ff}, as shown in Eq. (1.47).

$$M_{ff} = G_{ff}\delta \tag{1.47}$$

The feedforward control gain depends on the designed target side slip angle and yaw rate. In the early stage of Nagai's study, target yaw rate was set as the function as shown in Eq. (1.45) [91–93], while the target side slip angle was designed as first order inertia transfer function for steering angle of front wheel [42, 88, 89], as shown in Eq. (1.48).

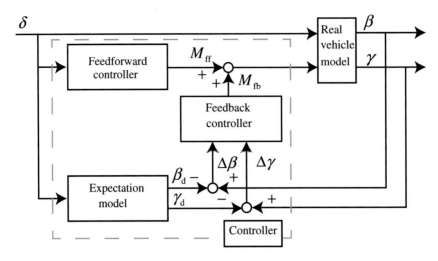

Fig. 1.1 Combined feedforward and feedback control structure

$$\beta_{\mathrm{d}} = \frac{K_\beta}{1 + \tau_\beta s}\delta \tag{1.48}$$

wherein, K_β refers to the amplification coefficient and τ_β refers to the time constant. However, through his subsequent study, it was discovered that the target side slip angle can be designed as 0 since it is generally very small. Therefore, the control side slip angle was set as 0 [30, 43, 90–93] in his subsequent study as well as others' study [6, 28, 94, 95]. In this case, the feedforward response coefficient can be designed as follows according to Eq. (1.37):

$$G_{\mathrm{ff}} = \frac{a_{22}b_1 - a_{12}b_2}{a_{12}e_2} \tag{1.49}$$

The feedback control is equivalent to the correction to external disturbance, with the input generally being yaw rate error $\gamma - \gamma_{\mathrm{d}}$ and side slip angle error $\beta - \beta_{\mathrm{d}}$, and output being feedback control moment M_{fb}. There are many methods for calculating feedback control moment, including the typical LQR control [30, 42, 43, 88–90], generalized prediction control [6, 96], sliding-mode control [28, 97], etc. Currently, the most common feedback control method is LQR control, which is of good stability and robustness and has passed the verification by real vehicles. Through superposition of feedforward control and feedback control, actually expected direct yaw moment $M_{z,\mathrm{d}}$ can be obtained, as shown in Eq. (1.50).

$$M_{z,\mathrm{d}} = M_{\mathrm{ff}} + M_{\mathrm{fb}} \tag{1.50}$$

1.3.2 Driving Force Control Allocation

The driving force control allocation involves the comprehensive realization of longitudinal and lateral control targets through multi-wheel driving force control allocation. Distributed electric vehicles are equipped with multiple wheels of independent drive, but generally the control target only covers expected longitudinal driving force $F_{x,d}$ and direct yaw moment $M_{z,d}$; therefore, it is a typical configuration for redundancy execution mechanism, which is subject to comprehensive design in terms of driving force control allocation.

1.3.2.1   Target and constraint conditions of desired control allocation

The desired full-vehicle longitudinal control target for distributed electric vehicles can be expressed by Eq. (1.51).

$$F_{1,x} + F_{2,x} + F_{3,x} + F_{4,x} = F_{x,d} \tag{1.51}$$

Different from full-vehicle resultant yaw moment (requiring consideration of the impact of lateral force), direct yaw moment is generated from the difference in longitudinal driving force on the two sides of vehicle; therefore, full-vehicle lateral desired control target of distributed electric vehicles can be expressed by Eq. (1.52).

$$\frac{b_f}{2}(F_{2,x} - F_{1,x}) + \frac{b_r}{2}(F_{4,x} - F_{3,x}) = M_{z,d} \tag{1.52}$$

The main purpose of control allocation is to satisfy drivers' demand for longitudinal drive and lateral stability through execution of longitudinal and lateral control. Moreover, it is generally necessary to meet certain constraint conditions during control allocation. According to the existing literatures, major constraint conditions in terms of drive motors are as follows.

(1) Constraint of motor execution performance
It generally refers to the maximum driving force constraint generated by a motor, which depends on the property of motor itself, as well as on the current power supply equipment, vehicle condition and tire condition. The Refs. [6, 28, 78, 82, 98] took the constraint of motor execution performance into consideration in driving force control allocation, as shown in Eq. (1.53).

$$|F_{i,x}| \le F_{i,\max} \tag{1.53}$$

(2) Constraint of motor failure
The actuator failure control has been widely applied in many fields currently, such as aviation and submarine [99–101]. The constraint of motor failure is frequently

realized through setting failure factors, which is adopted by the Refs. [82, 102, 103]—as shown in Eq. (1.54).

$$|F_{i,x}| \leq \xi F_{i,\max} \tag{1.54}$$

(3) Constraint of road surface adhesion

The longitudinal and lateral force of tires needs to be restrained within the adhesion circle [29]; otherwise, the expected longitudinal and lateral will not be executed effectively. The Refs. [28, 78, 86, 104] employed the constraint of road surface adhesion, as shown in Eq. (1.55).

$$|F_{i,x}| \leq \sqrt{\mu^2 F_{i,z}^2 - F_{i,y}^2} \tag{1.55}$$

(4) Constraint of drive anti-trackslip

Slipped tires may occur to distributed electric vehicles on road surface with low adhesion coefficient. In order to prevent over-trackslip of tires, drive anti-trackslip control shall be conducted to restrain the driving force generated by driving wheels [98, 105].

1.3.2.2 Driving Force Allocation Method Based on Rules

Driving force allocation based on certain rules is a good method for achieving sound robustness, with which reliable effect may be realized in terms of real vehicles. The Ref. [74] adopted equal distribution control method. First, it is necessary to set the total desired longitudinal driving moment $T_{\text{acc}}(T_{\text{acc}} = F_x R)$, difference value of desired yaw moment $T_{\text{ybf}}(T_{\text{ybf}} = M_z b/R)$, front-rear axle transfer moment T_{dnfr} during acceleration, front axle right-left transfer moment T_{dnf} during drive trackslip, and rear axle right-left transfer moment T_{dnr} during drive trackslip and then to allocate the forces to corresponding driving wheels, as shown in Eq. (1.56). This typical allocation method, though considering the constraint of drive anti-trackslip, lacks the involvement of other constraints, which may lead to failure in the moment execution.

$$\begin{cases} T_{\text{FL}} = \dfrac{1}{4}T_{\text{acc}} - \dfrac{1}{2}T_{\text{dnfr}} - \dfrac{1}{4}T_{\text{ybf}} + T_{\text{dnf}} \\[2mm] T_{\text{FR}} = \dfrac{1}{4}T_{\text{acc}} - \dfrac{1}{2}T_{\text{dnfr}} + \dfrac{1}{4}T_{\text{ybf}} - T_{\text{dnf}} \\[2mm] T_{\text{RL}} = \dfrac{1}{4}T_{\text{acc}} + \dfrac{1}{2}T_{\text{dnfr}} - \dfrac{1}{4}T_{\text{ybf}} + T_{\text{dnr}} \\[2mm] T_{\text{RR}} = \dfrac{1}{4}T_{\text{acc}} + \dfrac{1}{2}T_{\text{dnfr}} + \dfrac{1}{4}T_{\text{ybf}} - T_{\text{dnr}} \end{cases} \tag{1.56}$$

The Ref. [98] applied the control allocation method based on rules in the over-trackslip working condition. Under the condition of single-wheel trackslip or multi-wheel simultaneous trackslip, the driving moment of trackslip wheels is transferred to the non-trackslip wheels, thus satisfying the longitudinal driving performance and lateral stabilization performance of full vehicle; however, this method is only applicable to the over-trackslip working condition, without comprehensive solution to the drive failure, drive trackslip and control over direct yaw moment.

The Ref. [106] offered a basic principle for front/rear wheel drive regulation based on economical efficiency. According to this principle, different driving modes (front drive/four-wheel drive) shall be adopted based on different working conditions, in order to enhance economical efficiency: front drive shall be adopted in case of normal driving, while four-wheel drive may be adopted in case of starting, front-wheel trackslip, steering, deceleration and braking working condition, aiming at energy recovery as much as possible.

The Refs. [102, 107, 108] provided the logic for treatment of motor failure: the driving wheel in failure and driving wheel on opposite side will be disabled simultaneously in case of motor drive failure. With such treatment, part of driving performance can be maintained in case of single-wheel failure or co-axial double-wheel failure, but real-time moment allocation can not be realized according to the current vehicle state, partial motor failure is not taken into account, and motor response speed is not sufficiently considered, leading to weakened vehicle longitudinal drive performance and inadequate solution to reasonable driving in case of failure.

The Ref. [109] took the feature of redundancy driving source of distributed electric vehicles into consideration, and realized coordinated control over multiple vehicle wheels based on rules in case of single-wheel or double-wheel drive failure, thus satisfying drivers' demand for longitudinal and lateral control to a feasible extent. Although coordinated control is realized in case of complete failure of driving wheels, the conditions of partial driving wheel failure and over-trackslip are requiring further discussion.

1.3.2.3 Driving Force Allocation Method Based on Optimization

The redundancy configuration of distributed electric drive system makes it possible to optimize the driving force allocation, according to which scholars have put forward various optimization indexes and designed corresponding solutions to the optimization, including the following types.

(1) Driving force allocation mode based on economical efficiency
The Refs. [82, 102, 110, 111] optimized the economical efficiency under driving working condition by adopting the target shown in Eq. (1.57) in optimizing the driving moment of motors.

$$\min \frac{1}{\sum_{i=1}^{4} T_i} \sum_{i=1}^{4} \frac{T_i}{\varepsilon_i(T_i, n_i)} \tag{1.57}$$

The Ref. [111] optimized the economical efficiency under braking working condition by adopting the target shown in Eq. (1.58) in optimizing the braking moment of motors.

$$\min \frac{1}{\sum_{i=1}^{4} T_i} \sum_{i=1}^{4} T_i \varepsilon_i (T_i, n_i) \tag{1.58}$$

In Eqs. (1.57) and (1.58)—ε_i refers to the driving or braking efficiency under current moment and rotate velocity of motors.

(2) Driving force allocation mode based on surplus adhesion capacity
The Ref. [112] put forward the concept of tire load rate, specific definition of which is as shown in Eq. (1.59).

$$\varepsilon = \frac{\sqrt{F_x^2 + F_y^2}}{\mu F_z} \tag{1.59}$$

wherein, ε refers to the tire load rate; the smaller the ε is, the larger the acting force generated by current tires and the ground will be, and the better the vehicle safety will be. The Refs. [86, 112–119] applied the tire load in the optimized allocation of driving force, wherein the Refs. [86, 112, 113] adopted the optimization target as shown in Eq. (1.60).

$$\min \quad J = \sum_{i=1}^{4} D_i \varepsilon_i^2 \tag{1.60}$$

wherein (1.60)—D_i refers to the load rate weight coefficient of each wheel. The Refs. [6, 82, 96, 120] simplified the Eq. (1.60) to avoid the impact of lateral force on optimization effect, as shown in Eq. (1.61).

$$\min \quad J = \sum_{i=1}^{4} D_i \left(\frac{F_{i,x}}{\mu F_{i,z}} \right)^2 \tag{1.61}$$

The Refs. [95, 114–117] adopted the optimization target as shown in Eq. (1.62)—and the vehicle safety is ensured by minimizing the load rate of the wheel with maximum load rate.

$$\min \quad J = \max\{\varepsilon_1^2, \varepsilon_2^2, \varepsilon_3^2, \varepsilon_4^2\} \tag{1.62}$$

The Ref. [118] adopted the optimization target as shown in Eq. (1.63)—according to which uniform utilization of the load rate of each wheel is a better way for control allocation.

$$\min \quad J = \frac{1}{4} \sum_{i=1}^{4} \left(\varepsilon_i^2 - E(\varepsilon_i^2) \right)^2 \tag{1.63}$$

The Ref. [119] took an overall consideration of the concept in the Refs. [86, 112, 113, 118]—and put forward the optimization target as shown in Eq. (1.64). In this way, overall decrease of the sum of tire load rate variance and load rate for wheels, and uniform and sufficient utilization of wheel adhesive capacity and reduction of full-vehicle load rate are achieved, thus improving the manipulation stability of vehicles.

$$\min \quad J = \frac{1}{4} \sum_{i=1}^{4} \left(\varepsilon_i^2 - E(\varepsilon_i^2) \right)^2 + \lambda \sum_{i=1}^{4} \varepsilon_i^2 \tag{1.64}$$

wherein, λ refers to the weight coefficient of load rate sum of each wheel.

(3) Analytic method for optimized control
There are many analytic methods for optimized control, including analytical method [6, 86, 96, 112, 120] and various non-linear optimization methods (such as numerical method [111], complex method [110] and quadratic programming method [82, 103, 114, 115], among which it is noticeable in the Refs. [86, 112] that the desired longitudinal and lateral forces for wheels were obtained by virtue of the principle that the partial derivative of objective function at optimal point is 0, with the three constraints designed being target longitudinal force, target lateral force and target yaw moment and controllable variable being the four-wheel longitudinal and lateral forces, which is applicable to the electric automobiles with four-wheel steering and four-wheel drive.

1.3.3 Motor Discrepancy Compensation

The vehicle performance directly depends on the steady-state and dynamic response property of vehicle drive parts. As for vehicles with single-drive source, the motion conditions can be corrected through drivers' operation, even through the driving source features steady-state error or poor dynamic property; while as for distributed electric vehicles with multiple drive parts, the discrepancy in motor control performance will exert great impact on the lateral stability and returnability of vehicles, making it difficult in conducting coordinated control over driving force of wheels. Therefore, full-vehicle dynamic performance can only be ensured when each driving wheel bears equal moment response property, which is rightly the target of motor discrepancy compensation.

Currently, there are no adequate solutions to the moment discrepancy in distributed electric driving wheels; instead, most literatures involve the coordinated control over rotate velocity of motor driving wheels to ensure straight driving, which is common for crawler vehicles and engineering mechanical vehicles. Wherein, the Ref. [121] focused on the straight driving stability of double-motor independent crawler vehicles, and performed timely rotate velocity compensation in control command according to the displacement difference value of active wheels on both sides, which is accumulated by displacement compensation algorithm (only applicable to crawler vehicles) during straight driving, so as to reduce the rotate velocity difference value of motors on both sides and further reduce the off-tracking quantity during straight driving. The Ref. [122] performed step-by-step regulation of pressure of hydraulic pumps on both sides according to the rotate velocity difference value of crawler vehicle motors on both sides during straight driving.

The Refs. [123, 124] studied the impact of motor discrepancy on the straight driving performance of distributed electric vehicles, and designed feedforward compensation control aiming at the static error and proportional error and feedback compensation control in terms of dynamic error of motor system. A major disadvantage of this method was lack of consideration for dynamic response property of motors: the dynamic response effect of feedback compensation tracing target was not satisfactory and the integration of static error and dynamic error compensation was not considered.

1.4 Main Study Contents of This Dissertation

It can be seen from domestic and foreign researches that as a new type of vehicles, distributed electric vehicles are bound to be studied in terms of State Estimation and Coordinated Control, and a large number of key technological issues are to be discussed, covering the following aspects:

(1) The state estimation system for distributed electric vehicles has not been complete yet, and attention has only been paid to single-parameter estimation; in addition, the unique advantages of distributed electric vehicles in State Estimation have not been sufficiently utilized. By now, domestic and foreign researchers have had relatively adequate discussions on the vehicle motion state, tire acting force, vehicle intrinsic parameters and road surface parameters, except the relation among various state variables. Most researches fail to conduct overall design for state estimation system with only desired state parameters being observed. In this sense, it will be a major trend to conduct combined observation of various state to provide sound support for vehicle dynamic control; establishing clear and complete dynamic observation system and combined variable observation are also issues worthy of exploring.

(2) Current state estimation methods are generally based on traditional vehicles, without considering the intrinsic characteristics and multi-information sources of distributed electric vehicles. In fact, accurate dynamic information (driving wheel moment) and kinematic information (driving wheel rotate velocity) of distributed electric vehicles can be obtained from electric driving wheels without the need of more sensors. Therefore, methods for combined full-vehicle State Estimation should be developed by virtue of the multi-information sources of distributed electric vehicles.

(3) In terms of coordinated control for distributed electric vehicles, the set lateral control target usually can not fully reflect the demand of drivers. Most current researches focus on the tracing of multiple motion control targets (yaw rate and side slip angle) for vehicles, lacking comprehensive consideration of vehicle non-linear properties, self-adaptation regulation of control targets, and discussion on importance and switching method of control targets. Therefore, a control method should be designed for control target regulation according to current vehicle and surrounding conditions and real-time tracing of regulated targets after vehicles enter strong non-linear area so as to ensure stability of vehicles.

(4) In terms of coordinated control for distributed electric vehicles, most existing driving force control allocation methods fail to fully consider the allocation constraints and achieve complete performance of allocated driving force; moreover, effective design is lacked for failure control of electric drive system and moment coordination during anti-trackslip control, as the advantage of multi-execution units of Distributed Electric Vehicles is not sufficiently utilized.

(5) In terms of coordinated control for distributed electric vehicles, there is few research on motor discrepancy compensation at the moment. Ideally, feedback compensation is not necessary if electric driving wheels fully execute the desired moment set according to medium-layer allocation; actually, the discrepancy in motors becomes visible after long-term running, which will seriously impact the full-vehicle dynamic control effect without timely compensation control. Further study may be required on motor property compensation.

For the purpose of solving the above problems, this dissertation studies on the state estimation and coordinated control for distributed electric vehicles based on the theories on vehicle dynamics and control, and carries out system design, key technology and platform development herein.

(1) Structural design of State Estimation system and Coordinated Control system
The system structural design was completed according to the overall demand for State Estimation and Coordinated Control as well as characteristics of Distributed Electric Vehicles, making the combined observation of multiple observation state variables and the Coordinated Control through layered structure realized, and improving the dynamic performance of vehicles through modular design for each control link.

(2) Estimation methods for state parameters of Distributed Electric Vehicles
Combined observation of multiple vehicle states was conducted with existing common in-vehicle sensors, by virtue of the role of driving wheels in distributed electric vehicles as both information source and power source. In order to ensure observation effect, estimation of tire vertical force, mass of full vehicle and road surface gradient was performed before creating non-linear vehicle model and introducing unscented particle filtering method in the combined observation of multiple state parameters, including longitudinal velocity, side slip angle, yaw rate and tire lateral force. This method ensures sound timeliness and accuracy of observation and is of favorable effect.

(3) Setting of direct yaw moment of distributed electric vehicles
Through control over direct yaw moment, the stability of Distributed Electric Vehicles can be ensured, and the impact of non-linear vehicle system and multiple uncertain parameters on control effect can be avoided. This dissertation introduces the impact of non-linear factors, including vertical load transfer and change in road surface adhesion coefficient, based on double-track vehicle model, and adopts the method for control target weight regulation with $\beta - \dot{\beta}$ phase diagram to avoid vehicle instability caused by any change in road surface adhesion coefficient or high side slip angle velocity, thus maintaining the stability of vehicles to a larger extent.

(4) Driving force control allocation of distributed electric vehicles
In order to obtain the desired direct yaw moment and longitudinal driving force set for upper-layer control, constraint conditions for allocation of multiple driving forces were analyzed in a deep-through manner according to the vehicle motion state, motor working state and dynamic control demand in the process of control coordination, thus ensuring effective motor performance of allocated driving force and even certain driving capacity and vehicle stability in case of drive failure or trackslip. In case of any contradiction between demand targets (longitudinal driving force and direct yaw moment), dynamic self-adaptation weight regulation was applied to specific demand targets, making it possible to meet major demand targets in priority. This dissertation puts forward the concept of motor usage rate for equal utilization of driving wheels and protection of failed wheels. Through the design of objective function, the driving force allocation for driving wheels was optimized by non-linear method, making uniform driving force allocation among motors realized, satisfying the longitudinal and lateral demand and achieving comprehensive solution to the motor failure, drive trackslip and control allocation.

(5) Motor property compensation for distributed electric vehicles
In order to execute the desired motor driving force set by control allocation layer, this dissertation puts forward the control method for motor property self-adaptation compensation, through which motor dynamic and steady-state response was analyzed in a deep-through manner, motor response properties and random error caused by external disturbance were considered, and self-adaptation control rate was set by

Lyapunov method, ensuring sound execution of motor driving force under various working conditions and improving the dynamic control effect.

(6) Simulation verification of state estimation and coordinated control
For more accurate verification of the developed State Estimation and Coordinated Control, this dissertation establishes a combined CarSim-Simulink simulation test platform according to the new features of Distributed Electric Vehicles, based on which simulation research on State Estimation and Coordinated Control was performed, verifying the accuracy and reasonability of the developed systems for state estimation and coordinated control.

(7) Experimental verification of systems for state estimation and coordinated control
The effectiveness and applicability of systems for State Estimation and Coordinated Control for Distributed Electric Vehicles during actual driving was verified through corresponding experimental plan prepared according to the rapid control prototype (RCP), which was developed based on the algorithms for state estimation and coordinated control through simulation verification.

References

1. Yu Z, Gao X (2009) Review of vehicle state estimation problem under driving situation. J Mech Eng 45(5):20–33
2. Carlson C R (2004) Estimation with application for automobile dead reckoning and control. Stanford University, USA
3. Bevly DM, Gerdes JC, Wilson C (2002) Use of GPS based velocity measurements for measurement of sideslip and wheel slip. Veh Syst Dyn 38(2):127–147
4. Ryu J (2004) State and parameter estimation for vehicle dynamics control using GPS. Stanford University, USA
5. Schoepflin TN, Dailey DJ (2003) Dynamic camera calibration of roadside traffic management cameras for vehicle speed estimation. IEEE Trans Intell Transp Syst 4(2):90–98
6. Liu L, Luo Y, Li K (2009) Observation of road surface adhesion coefficient based on normalized tire model. J Tsinghua Univ (Nat Sci) 49(5):116–120
7. Kiencke U, Nielsen L (2010) Automotive control systems: for engine, driveline, and vehicle, 2nd edn. Springer, New York
8. Qi Z, Zhang J (2003) Study on reference vehicle velocity determination for ABS based on vehicle ABS/ASR/ACC integrated system. J Automot Eng 25(6):617–620
9. Jiang F, Gao Z (2000) An adaptive nonlinear filter approach to the vehicle velocity estimation for ABS. In: Proceedings of the international conference on control applications, Anchorage, Alaska, USA, Sept 2000, pp 490–495
10. Kalman RE (1960) A new approach to linear filtering and prediction problems. Trans ASME J Basic Eng 82(Series D):35–45
11. Welch G, Bishop G (2006) An introduction to the Kalman filter. Technical Report, Department of Computer Science, University of North Carolina at Chapel Hill. http://www.cs.unc.edu/~welch/kalman/
12. Lee H (2006) Reliability indexed sensor fusion and its application to vehicle velocity estimation. J Dyn Syst Meas Contr 128(2):236–243

13. Kobayashi K, Cheok KC, Watanabe K (1995) Estimation of absolute vehicle speed using fuzzy logic rule-based Kalman filter. In: Proceedings of the American control conference, Seattle, WA, USA, June 1995, pp 3086–3090

14. Watanabe K, Kinase K, Kobayashi K (1993) Absolute speed measurement of vehicles from noisy acceleration and erroneous wheel speed. In: Proceedings of the intelligent vehicles symposium, Tokyo, Japan, pp 271–276

15. Chu W, Li S, Jiang Q et al (2011) Estimation of full-wheel independently-driven electric vehicle based on multi-information integration. J Autom Eng 33(11):962–966

16. Imsland L, Johansen TA, Fossen TI et al (2006) Vehicle velocity estimation using nonlinear observers. Automatica 42(12):2091–2103

17. Shraim H, Ananou B, Fridman L, et al (2006) Sliding mode observers for the estimation of vehicle parameters, forces and states of the center of gravity. In: Proceedings of the IEEE conference on decision & control, San Diego, California, USA, Dec 2006, pp 1635–1640

18. Cherouat H, Braci M, Diop S (2005) Vehicle velocity, side slip angles and yaw rate estimation. In: Proceedings of the IEEE international symposium on industrial electronics, Dubrovnik, Croatia, June 2005, pp 349–354

19. Xu T, Gao X (2007) Overview of Development Trend of Vehicle Longitudinal Velocity Estimation Algorithm. Shanghai Auto 6:39–42

20. Piyabongkarn D, Rajamani R, Grogg JA et al (2009) Development and experimental evaluation of a slip angle estimator for vehicle stability control. IEEE Trans Control Syst Technol 17(1):78–88

21. van Zanten AT (2000) Bosch ESP systems: 5 Years of experience. SAE technical paper: 2000–01–1633

22. Aoki Y, Uchida T, Hori Y (2005) Experimental demonstration of body slip angle control based on a novel linear observer for electric vehicle. In: Proceedings of the IEEE conference on industrial electronics society, Raleigh, North Carolina, USA, Nov 2005, pp 2620–2625

23. Venhovens PJT, Naab K (1999) Vehicle dynamics estimation using Kalman filters. Veh Syst Dyn 32(2):171–184

24. Farrelly J, Wellstead P (1996) Estimation of vehicle lateral velocity. In: Proceedings of the IEEE conference on control applications, Dearborn, Michigan, USA, Sept 1996, pp 552–557

25. Geng C, Mostefai L, Denaï M et al (2009) Direct yaw-moment control of an in-wheel-motored electric vehicle based on body slip angle fuzzy observer. IEEE Trans Ind Electron 56(5):1411–1419

26. Unger I, Isermann R (2006) Fault tolerant sensors for vehicle dynamics control. In: Proceedings of the American control conference, Minneapolis, Minnesota, USA, June 2006, pp 3948–3953

27. Chumsamutr R, Fujioka T, Abe M (2006) Sensitivity analysis of side-slip angle observer based on a tire model. Veh Syst Dyn 44(7):513–527

28. Zou G (2008) Integral dynamical control of four-wheel independently-driven electric vehicle system. Tsinghua University, Beijing

29. Yu Z (2009) Automobile theory, No.5 version. Tsinghua University Press, Beijing

30. Nagai M (2007) The perspectives of research for enhancing active safety based on advanced control technology. Veh Syst Dyn 45(5):413–431

31. Stphant J, Charara A, Meiz D (2004) Virtual sensor: application to vehicle sideslip angle and transversal forces. IEEE Trans Ind Electron 51(2):278–289

32. Stphant J, Charara A (2005) Observability matrix and parameter identification application to vehicle tire cornering stiffness. In: Proceedings of the European control conference, Sevile, Spain, Dec 2005, pp 6734–6739

33. Hiemer M, von Vietinghoff A, Kiencke U (2005) Determination of the vehicle body side slip angle with non-linear observer strategies. SAE technical paper: 2005–01–0400

34. Doumiati M, Victorino A, Charara A, et al (2009) Estimation of vehicle lateral tire-road forces: a comparison between extended and unscented filtering. In: Proceedings of the European control conference, Budapest, Hungary, Aug 2009, pp 4804–4809

35. Gao X, Yu Z, Chen X (2009) Model based yaw rate estimation of electric vehicle with 4 in-wheel motors. SAE technical paper: 2009–01–0463
36. Hac A, Simpson M (2000) Estimation of vehicle side slip angle and yaw rate. SAE technical paper: 2000–01–0696
37. Gao Y, Gao Z, Li X (2005) Vehicle yaw velocity soft measurement algorithm based on adaptive Kalman filter method. J Jiangsu Univ (Nat Sci) 26(1):24–27
38. Dugoff H, Fancher PS, Segel L (1970) The Influence of lateral load transfer on directional response. SAE technical paper 700377
39. Liu Q, Guo K, Chen B (2000) Analysis on tire brush model. J Agric Mach 31(1):19–22
40. Pacejka HB (2012) Tire Veh Dyn, 3rd edn. Elsevier, Oxford
41. Dugoff H, Fancher PS, Segel L (1969) Tire performance characteristics affecting vehicle response to steering and braking control inputs. Final Report. Technical Report, Office of Vehicle Systems Research, US National Bureau of Standards. http://hdl.handle.net/2027.42/1387
42. Nagai M, Yamanaka S, Hirano Y (1996) Integrated control law of active rear steering control. In: Proceedings of the international symposium on advanced vehicle control. Aachen, Germany, July 1996, pp 451–469
43. Nagai M, Shino M, Gao F (2002) Study on integrated control of active front steer angle and direct yaw moment. JSAE Rev 23(3):309C315
44. Antonov S, Fehn A, Kugi A (2011) Unscented Kalman filter for vehicle state estimation. Veh Syst Dyn 49(9):1497C1520
45. Reif K, Renner K, Saeger M (2008) Using the unscented Kalman filter and a non-linear two-track model for vehicle state estimation. In: Proceedings of the international federation of automatic control. Seoul, Korea, July 2008, pp 8570–8575
46. Yu X (2010) Study on integral control strategy of vehicle active suspension. Jilin University, Changchun
47. Schofield B, Hägglund T (2008) Optimal control allocation in vehicle dynamics control for rollover mitigation. In: Proceedings of the American control conference, Seattle, Washington, USA, June 2008, pp 3231–3236
48. Hyun D, Langari R (2003) Modeling to predict rollover threat of tractor-semitrailers. Veh Syst Dyn 39(6):401–414
49. Schofield B (2006) Vehicle dynamics control for rollover prevention. Lund University, Sweden
50. Nam K, Oh S, Fujimoto H, et al (2011) Vehicle state estimation for advanced vehicle motion control using novel lateral tire force sensors. In: Proceedings of the American control conference, San Francisco, California, USA, July 2011, pp 5372–5377
51. Cho W, Yoon J, Yim S et al (2010) Estimation of tire forces for application to vehicle stability control. IEEE Trans Veh Technol 59(2):638–649
52. Yoon J, Yi K (2006) A rollover mitigation control scheme based on rollover index. In: Proceedings of the American control conference, Minneapolis, Minnesota, USA, June 2006, pp 4853–4858
53. Doumiati M, Victorino A, Charara A et al (2008) An estimation process for vehicle wheel-ground contact normal forces. In: Proceedings of the world congress of the international federation of automatic control, Seoul, Korea, July 2008, pp 7110–7115
54. Bae HS, Gerdes JC (2000) Parameter estimation and command modification for longitudinal control of heavy vehicles. In: Proceedings of the international symposium on advanced vehicle control, Ann Arbor, Michigan, USA
55. Bae HS, Ryu J, Gerdes JC (2001) Road grade and vehicle parameter estimation for longitudinal control using GPS. In: Proceedings of the IEEE conference on intelligent transportation systems. Oakland, California, USA, Aug 2001
56. Johansson K (2005) Road slope estimation with standard truck sensors. KTH Royal Institute of Technology, Sweden
57. Jansson H, Kozica E, Sahlholm P et al (2006) Improved road grade estimation using sensor fusion. In: Proceedings of the 12th Reglermöte, Stockholm, Sweden, May 2006

58. Sahlholm P, Johansson KH (2010) Road grade estimation for look-ahead vehicle control using multiple measurement runs. Control Eng Pract 18(11):1328–1341
59. Parviainen J, Hautamäki J, Collin J et al (2009) Barometer-aided road grade estimation. Proceedings of the world congress of the international association of institutes of navigation. Stockholm, Sweden, Oct 2009
60. Zhang T (2010) Behavior matching of vehicle driving roads. Tsinghua University, Beijing
61. Zhang T, Yang D, Li T et al (2010) Vehicle state estimation system aided by inertial sensors in GPS navigation. In: Proceedings of the international conference on electrical and control engineering, Wuhan, China, June 2010
62. Grewal M, Weill L, Andrewsa A (2007) Global positioning systems, inertial navigation, and integration. Wiley, Hoboken
63. Parkum JE, Poulsen NK, Holst J (1992) Recursive forgetting algorithms. Int J Control 55(1):109–128
64. Saelid S, Foss B (1983) Adaptive controllers with a vector variable forgetting factor. In: Proceedings of the IEEE conference on decision and control, San Antonio, TX, USA, Dec 1983, pp 1488–1494
65. Vahidi A, Druzhinina A, Stefanopoulou A et al (2003) Simultaneous mass and time-varying grade estimation for heavy-duty vehicles. In: Proceedings of the American control conference, Denver, Colorado, USA, June 2003, pp 4951–4956
66. Vahidi A, Stefanopoulou A, Peng H (2003) Experiments for online estimation of heavy vehicles mass and time-varying road grade. In: Proceedings of the ASME international mechanical engineering congress and exposition, Washington, DC, USA
67. Vahidi A, Stefanopoulou A, Peng H (2005) Recursive least squares with forgetting for online estimation of vehicle mass and road grade: theory and experiments. Veh Syst Dyn 43(1):57–75
68. McIntyre ML, Ghotikar TJ, Vahidi A et al (2009) A two-stage lyapunov-based estimator for estimation of vehicle mass and road grade. IEEE Trans Veh Technol 58(7):3177–3185
69. Sastry S, Bodson M (2011) Adaptive control: stability, convergence and robustness. Dover Publications, Mineola
70. Holm EJ (2011) Vehicle mass and road grade estimation using Kalman filter. Linköping University, Sweden
71. Lingman P, Schmidtbauer B (2002) Road slope and vehicle mass estimation using Kalman filtering. Veh Syst Dyn 37(Supplement):12C–23
72. Winstead V, Kolmanovsky IV (2005) Estimation of road grade and vehicle mass via model predictive control. In: Proceedings of the IEEE conference on control applications, Toronto, Canada, Aug 2005
73. Eriksson A (2009) Implementation and evaluation of a mass estimation algorithm. KTH Royal Institute of Technology, Sweden
74. Kamachi M, Walters K (2006) A research of direct yaw-moment control on slippery road for in-wheel motor vehicle. In: Proceedings of the international battery, hybrid and fuel cell electric vehicle symposium & exposition, Yokohama, Japan, Oct 2006, pp 2122–2133
75. Raksincharoensak P, Nagai M (2006) Adaptive direct yaw moment control based on driver intention. Proceedings of the international symposium on advanced vehicle control, Taipei, China, Aug 2006, pp 753–758
76. Tahami F, Farhangi S, Kazemi R (2004) A fuzzy logic direct yaw-moment control system for all-wheel-drive electric vehicles. Veh Syst Dyn 41(3):203–221
77. Kim J, Kim H (2007) Electric vehicle yaw rate control using independent in-wheel motor. In: Proceedings of the power conversion conference, Nagoya, Japan, April 2007, pp 705–710
78. Li D (2008) Study on vehicle dynamical integrated control based on optimal allocation. Shanghai Jiaotong University, Shanghai
79. Li L, Song J, Wang H et al (2006) Linear subsystem model for real-time control of vehicle stability control system. In: Proceedings of the IEEE conference on robotics, automation and mechatronics, Bangkok, Thailand, Dec 2006
80. Abe M, Kano Y, Suzuki K et al (2001) Side-slip control to stabilize vehicle lateral motion by direct yaw moment. JSAE Rev 22(4):413–419

81. Boada BL, Boada MJL, Dłaz V (2005) Fuzzy-logic applied to yaw moment control for vehicle stability. Veh Syst Dyn 43(10):753–770

82. Wang B (2009) Study on experiment platform and driving force control system for four-wheel independently-driven electric vehicle. Tsinghua University, Beijing

83. Pi D, Chen N, Wang J (2008) Application of fuzzy logics in vehicle stability control system. J Southeast Univ (Nat Sci) 38(1):43–48

84. Kim D, Hwang S, Kim H (2010) Vehicle stability enhancement of four-wheel-drive hybrid electric vehicle using rear motor control. IEEE Trans Veh Technol 57(2):727–735

85. Park JH, Ahn WS (1999) H_∞ yaw-moment control with brakes for improving driving performance and stability. In: Proceedings of the international conference on advanced intelligent mechatronics, Atlanta, Georgia, USA, Sept 1999, pp 747–752

86. Mokhiamar O, Abe M (2006) How the four wheels should share forces in an optimum cooperative chassis control. Control Eng Pract 14(3):295–304

87. Yih P, Gerdes JC (2005) Modification of vehicle handling characteristics via steer-by-wire. IEEE Trans Control Syst Technol 13(6):965–976

88. Nagai M, Hirano Y, Yamanaka S (1997) Integrated control of active rear wheel steering and direct yaw moment control. Veh Syst Dyn 27(5):357–370

89. Nagai M, Hirano Y, Yamanaka S (1998) Integrated robust control of active rear wheel steering and direct yaw moment control. Veh Syst Dyn 28(Supplement):416–421

90. Shino M, Miyamoto N, Wang Y et al (2000) Traction control of electric vehicles considering vehicle stability. In: Proceedings of the international workshop on advanced motion control. Nagoya, Japan, April 2000, pp 311–316

91. Shino M, Nagai M (2001) Yaw-moment control of electric vehicle for improving handling and stability. JSAE Rev 22(4):473–480

92. Shino M, Nagai M (2003) Independent wheel torque control of small-scale electric vehicle. JSAE Rev 24(4):449–456

93. Shino M, Raksincharoensak P, Kamata M et al (2004) Slide slip control of small-scale electric vehicle by DYC. Veh Syst Dyn 41(Supplement):487–496

94. Xiong L, Yu Z, Wang Y et al (2012) Vehicle dynamics control of four in-wheel motor drive electric vehicle using gain scheduling based on tire cornering stiffness estimation. Veh Syst Dyn 50(6):831–846

95. Yu Z, Jiang W, Zhang L (2008) Torque allocation control over four-wheel hub motor-driven electric vehicle. J Tongji Univ (Nat Sci) 36(8):1115–1119

96. Liu L, Luo Y, Jiang Q et al (2011) AFS/DYC chassis integrated control based on generalized prediction theory. J Autom Eng 33(1):52–56

97. Zou G, Luo Y, Li K et al (2008) Improvement maneuverability and stability of independent 4WD EV by DYC based on dynamic regulation of control target. J Mech Syst Transp Logistic 1(3):305–318

98. Chu W, Luo Y, Zhao F et al (2012) Driving force coordinated control of distributed electric drive vehicle. J Autom Eng 34(3):185–189

99. Harkegard O (2003) Backstepping and control allocation with applications to flight control. LinkOping University, Sweden

100. Omerdic E, Roberts G (2004) Thruster fault diagnosis and accommodation for open-frame underwater vehicles. Control Eng Prac 12(12):1575–1598

101. Podder TK, Sarkar N (2001) Fault-tolerant control of an autonomous underwater vehicle under thruster redundancy. Robot Auton Syst 34(1):39–52

102. Bo W, Yugong L, Fan J et al (2010) Driving force allocation algorithm of four- wheel independently-driven electric vehicle based on control allocation. J Autom Eng 32(2):128–132

103. Chu W, Luo Y, Dai Y et al (2012) Traction fault accommodation system for four wheel independently driven electric vehicle. In: Proceedings of the international battery, hybrid and fuel cell electric vehicle symposium & exposition, Los Angeles, CA, USA, May 2012

104. Liu L, Luo Y, Li K (2010) Study on vehicle all-wheel longitudinal force allocation based on dynamic target control. J Autom Eng 32(1):60–64

105. Hu J, Yin D, Hori Y (2011) Fault-tolerant traction control of electric vehicles. Control Eng Pract 19(2):204–213
106. Kondo K, Sekiguchi S, Tsuchida M (2002) Development of an electrical 4WD system for hybrid vehicles. SAE technical paper: 2002–01–1043
107. Mutoh N, Takahashi Y, Tomita Y (2008) Failsafe drive performance of FRID electric vehicles with the structure driven by the front and rear wheels independently. IEEE Trans Ind Electron 55(6):2306–2315
108. Mutoh N (2009) Front-and-rear-wheel-independent-drive type electric vehicle (FRID EV) with the outstanding driving performance suitable for next-generation adavanced EVs. In: Proceedings of the vehicle power and propulsion conference, Dearborn, Michigan, USA, Sept 2009, pp 1064–1070
109. Chu W, Luo Y, Han Y et al (2012) Failure control over distributed electric drive vehicle based on rules. J Mech Eng 48(10):90–95
110. Fan J, Mao M (2007) Study on driving force allocation strategy of three-axle electric drive vehicle based on economy. Veh & Power Technol 1:49–51
111. Yu Z, Zhang L, Xiong L (2005) Optimal torque allocation control based on economy improvement for four-wheel electric drive vehicle. J Tongji Univ (Nat Sci) 33(10):1115–1119
112. Abe M, Mokhiamar O (2004) Effects of an optimum cooperative chassis control from the viewpoint of tire workload. Trans Soc Autom Eng Jpn 35(3):215–221
113. Mokhiamar O, Abe M (2004) Simultaneous optimal distribution of lateral and longitudinal tire forces for the model following control. J Dyn Syst Meas Control 126(4):753–763
114. Nishihara O, Kumamoto H (2006) Minimax optimizations of tire workload exploiting complementarities between independent steering and traction/braking force distributions. In: Proceedings of the international symposium on advanced vehicle control. Taipei, China, Aug 2006, pp 713–718
115. Nishihara O, Hiraoka T, Kumamoto H (2006) Optimization of lateral and driving/braking force distribution of independent steering vehicle (minimax optimization of tire workload). Trans Soc Autom Eng Jpn 72(714):537–544
116. Ono E, Hattori Y, Muragishi Y et al (2006) Vehicle dynamics integrated control for four-wheel-distributed steering and four-wheel-distributed traction/braking systems. Veh Syst Dyn 44(2):139–151
117. He P, Hori Y (2006) Improvement of EV maneuverability and safety by disturbance observer based dynamic force distribution. In: Proceedings of the international battery, hybrid and fuel cell electric vehicle symposium & exposition. Yokohama, Japan, Oct 2006, pp 1818–1827
118. Peng H, Sabahi R, Chen S et al (November 2011) Integrated vehicle control based on tire force reserve optimization concept. In: Proceedings of the ASME international mechanical engineering congress and exposition, Denver, Colorado, USA, Nov 2011
119. Dai Y, Luo Y, Chu W et al (2012) Optimum tire force distribution for four-wheel-independent drive electric vehicle with active front steering. In: Proceedings of the international symposium on advanced vehicle control, Seoul, Korea, Sept 2011
120. Zou G, Luo Y, Li K (2009) Optimal method for all-wheel longitudinal force allocation of four-wheel independently-driven electric vehicle. J Tsinghua Univ (Nat Sci) 49(5):719–722
121. Sun H, Zhang C, Cao L (2005) Study on straight driving stability control over double-motor independently-driven crawler vehicle. Veh Pow Technol 4:26–29
122. Wang X, Zhou X (2007) Study on straight driving correction method for full hydraulic bull-dozer. J Mech Eng 38(2):17–21
123. Zhang H, Wang Q, Jin L (2007) Study on the straight-line running stability of the four-wheel independent driving electric vehicles. SAE technical paper: 2007–01–3488
124. Zhang H (2009) Study on torque coordinated control over electric vehicle driven by electric wheel. Jilin University, Changchun

Chapter 2
State Estimation and Coordinated Control System

Abstract By taking full advantage of distributed electric vehicles in vehicle dynamic control field, optimized design has been made on dynamic control of full vehicle based on state estimation and coordinated control system of distributed electric vehicles. For tackling such problems concerning distributed electric vehicles as imperfect system in state estimation field, insufficient coordination degree in coordinated control field, poor control adaptability and poor disturbance-resistance capability, a system for dynamic state estimation and coordinated control of full vehicle by making full use of property of distributed electric vehicles is set, comprehensively improving precision of state estimation and dynamic performance of vehicles. This chapter studied state estimation and coordinated control system of distributed electric vehicles. Based on the concept of hierarchical design, the integrated system framework was designed, functions of various subsystems were specified, input and output of various subsystems were defined, logical relation of each layer was combed and technological difficulties and key points concerning the system was planned.

2.1 Structural Design of the System

Given the fact that state estimation and coordinated control system of distributed electric vehicles are complicated with multiple input and output being involved and that information and variables of variety are interconnected, centralized control tends to make the controller structure complex, functions ambiguous and development difficult and lead to difficulty in software maintenance and poor expansion performance. However, hierarchical control provides a more satisfactory solution for the above problem with superior expansion performance, making the control structure clear and conducive to development and maintenance. Therefore, hierarchical control structure was adopted in structure design of the control system in this dissertation.

Major functions to be realized: estimation of dynamic state parameter of vehicles according to the current operation information of drivers, feedback information of distributed electric driving wheels, INS information and GPS information; setting of expected longitudinal and lateral control targets by making use of the collected information of vehicle operating state parameters; calculation of allocation value of

© Springer-Verlag Berlin Heidelberg 2016
W. Chu, *State Estimation and Coordinated Control for Distributed Electric Vehicles*, Springer Theses, DOI 10.1007/978-3-662-48708-2_2

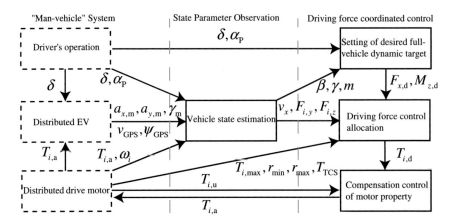

Fig. 2.1 Overall system of state estimation and coordinated control

driving force for each wheel by adopting proper control and allocation method; off-setting motor difference resulting from dynamic response and steady-state response of motor by adopting appropriate motor control method so as to execute the expected driving force allocation value for each wheel and realize longitudinal and lateral control targets by use of the allocation value.

After the above system functions were confirmed, overall structure of state estimation and coordinated control system of distributed electric vehicles was designed. As shown in Fig. 2.1, corresponding hierarchical control module was built in this dissertation based on the designed system function and logical relation of each layer was specified, comprehensively realizing the overall scheme for state estimation and coordinated control.

On the basis of in-vehicle INS information, GPS information, driver's operation information and electric driving wheel information, dynamic estimation system of distributed electric vehicles makes combined estimation on multiple important state variables of vehicles via certain state estimation methods, including longitudinal velocity v_x, side slip angle β, yaw rate γ, lateral force $F_{i,y}$ of tire, vertical force $F_{i,z}$ of tire, mass m and gradient θ of road surface. Acquisition of these state variables provides foundation for coordinated control of distributed electric vehicles. coordinated control system of distributed electric vehicles is of hierarchical structure, including upper layer, medium layer and lower layer. The upper layer control is used to determine dynamic demand of full vehicle and the expected control targets mainly based on driver's operation information and current state of vehicles estimated by vehicle state observer and compare the expected control target and actual control target, then outputting the expected longitudinal driving force $F_{x,d}$, and expected direct yaw moment $M_{z,d}$. Medium control allocates driving force, and its main target is to develop driving moment $T_{i,d}$ allocated by each driving wheel so as to satisfy dynamic demand $F_{x,d}$ and $M_{z,d}$ of full vehicle set by upper layer. In the process of control allocation, motion state of vehicles and running state of motor need to be considered in real time, including maximum in-wheel driving force constraint $T_{i,\max}$, moment slope constraint $r_{\max}$ and $r_{\min}$ and moment command T_{TCS} made by TCS. Lower control is used for motor property compensation and setting moment control command $T_{i,u}$ on the basis of expected driving force $T_{i,d}$ of each wheel and actual motor feedback moment $T_{i,a}$ set according to moment allocation of medium layer, making the actual motor output moment follow the expected moment.

The following section discusses state estimation and coordinated control of distributed electric vehicles.

(1) State estimation system of distributed electric vehicles
State estimation system is the foundation for vehicle dynamic control. Given the properties of distributed electric vehicles, this system observes various states of vehicles in a comprehensive way by making use of in-vehicle common sensor and in combination of feedback of distributed driving wheel, where available sensors include longitudinal acceleration sensor, lateral acceleration sensor, yaw rate sensor, roll rate sensor, steering wheel angle sensor, accelerator pedal sensor and GPS (global positioning system) and available feedback information of electric driving wheel include information of driving moment and rotate velocity of driving wheel. As shown in Fig. 2.2, this subsystem is designed by hierarchical structure. Firstly, various types of information that can be offered by distributed electric vehicles are collected and signals are preprocessed. "INS sensor integration system" can provide longitudinal

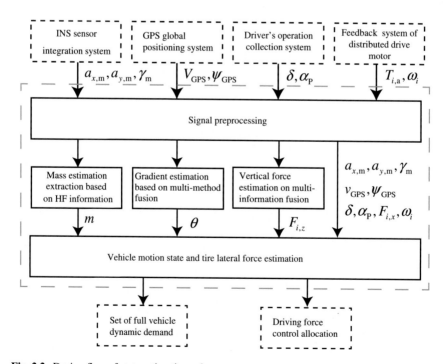

Fig. 2.2 Design flow of state estimation subsystem

acceleration $a_{x,\mathrm{m}}$, lateral acceleration $a_{y,\mathrm{m}}$ and yaw rate γ_m, while "GPS" can provide absolute velocity V_GPS and heading angle ψ_GPS of vehicles and "driver's operation information collection system" can steering angle δ of front wheel and accelerator pedal aperture α_P and "feedback information system of distributed drive motor" is able to offer actual moment $T_{i,\mathrm{a}}$ and rotate velocity ω_i of vehicles. Secondly, multiple parameters are estimated. By "mass estimation extraction based on high frequency information"; mass m of full vehicle can be obtained; gradient θ of road surface can be acquired according to "gradient estimation based on multi-method fusion"; vertical force $F_{i,z}$ of each wheel can be acquired by "vertical force estimation based on multi-information fusion". Ultimately, "vehicle motion state and tire lateral force estimation" is made by integrating information of multiple sensors and information of mass, gradient and vertical force acquired by estimation. By "vehicle motion state and tire lateral force estimation", longitudinal velocity v_x, side slip angle β, yaw rate γ and lateral force $F_{i,y}$ of each wheel of full vehicle can be estimated.

(2) Coordinated control system of distributed electric vehicles
Coordinated Control system of distributed electric vehicles is composed of the following three subsystems, including "determination of full vehicle dynamic demand", "driving force control allocation" and "compensation control of motor property".

1. Setting of desired dynamic target of full vehicle
Major function of "setting of desired dynamic target of full vehicle" subsystem is to calculate the desired longitudinal driving force $F_{x,\mathrm{d}}$ and direct yaw moment $M_{z,\mathrm{d}}$ of full vehicle according to driver's operation and running state of vehicles. Given that setting of longitudinal desired driving force has impact on dynamic performance of vehicles, the system makes open-loop control over the expected longitudinal driving force by making use of driver's operation information. Given that direct yaw moment has direct impact on stability of vehicles, this dissertation calculates the direct yaw moment of vehicles by utilizing the structure jointly controlled by feedforward and feedback where the feedforward develops the direct yaw moment M_ff of feedforward on the basis of expected targets and driver's operation information. Moreover, since vehicles are affected by a series of uncertain factors to different degrees in running process, feedback control is required to ensure lateral stability of vehicles and convergence of control targets. When actual targets deviate from the expected one, the feedback develops direct yaw moment M_fb and regulates the desired direct yaw moment and controls lateral motion state of full vehicle, guaranteeing stability of vehicles. Flow to develop desired dynamic target of full vehicle is as shown in Fig. 2.3.

2. Driving force control allocation
Major function of "driving force control allocation" is to allocate desired longitudinal driving force $F_{x,\mathrm{d}}$ and direct yaw moment $M_{z,\mathrm{d}}$ of full vehicle in "setting of desired dynamic target of full vehicle" subsystem among each driving wheel. Main purpose of control allocation is to satisfy longitudinal dynamic demand and lateral stability demand of full vehicle and takes full advantage of redundancy configuration structure of distributed electric vehicles, so as to optimize design of driving force

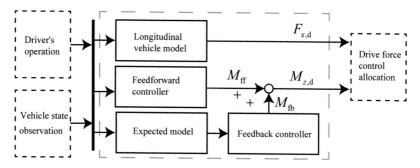

Fig. 2.3 Flow for setting of desired dynamic target of full vehicle

control allocation of full vehicle. Moment constraint of electric driving wheel needs to be taken into consideration for driving force control allocation, including execution capacity constraint of motor system, trackslip constraint, failure of driving wheel, etc. In addition, corresponding constraint functions are designed, including maximum in-wheel driving force constraint $T_{i,\mathrm{max}}$, moment slope constraint r_{max} and r_{min} and moment command T_{TCS} made by TCS. By such a way, upper limit $\bar{x}$ and lower limit $\underline{x}$ of driving force at the current moment can be acquired. Meanwhile, if desired longitudinal dynamic contracts to desired lateral stability, corresponding longitudinal and lateral regulation weighting function H_{F} and H_{M} is set to satisfy the current major desired targets of full vehicle, so as to realize self-adaptive control. Optimized design aims at controlling the usage rate of various motors and avoiding overuse of partial vehicle wheels. On such a basis, optimized target function is designed with constraint and expected driving moment $T_{i,\mathrm{d}}$ of each wheel is optimized and solved by non-linear optimization. Flow for driving force control allocation is as shown in Fig. 2.4.

3. Compensation control of motor property

Major function of "compensation control of motor property" subsystem is to execute target driving force of each set by control allocation layer in motor system, so as to satisfy the demand of dynamic control. In the process of compensation of motor property, reference model of motor is designed firstly by adopting self-adaptive control, and self-adaptive control law is designed on such a basis. Reference model is mainly used for simulating the expected motor response property. The system integrates target moment $T_{i,\mathrm{d}}$ set by control allocation layer of driving force, expected moment $T_{i,\mathrm{m}}$ set by reference model and the current actual moment $T_{i,\mathrm{a}}$, so as to make real-time dynamic regulation to control law. Self-adaptive controller calculates moment command $T_{i,\mathrm{u}}$ of motor in real time by making use of self-adaptive control rate and sends such a command to distributed drive motor ultimately so as to realize property compensation control of the motor, ensure actual moment of the motor following the expected one and meet the dynamic control demand of full vehicle. Flow for compensation control of motor property is as shown in Fig. 2.5.

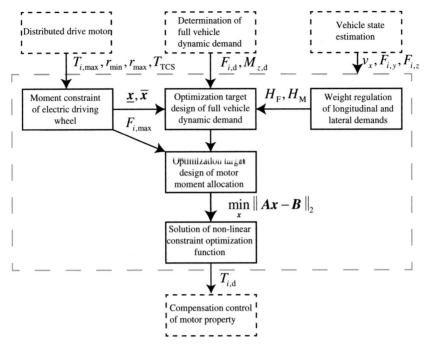

Fig. 2.4 Flow for driving force control allocation

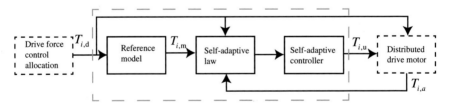

Fig. 2.5 Flow for compensation control of motor property

2.2 Technological Difficulties and Key Points

Technological difficulties and key points related to state estimation and coordinated control system of distributed drive vehicles fall into four parts, including design of state estimation system, design of desired dynamic target of full vehicle, driving force control allocation and compensation control of motor property. The following presents further description for each key technology mentioned.

(1) Design of state estimation system
The purpose of state estimation of distributed electric vehicles is to provide support for dynamic control of vehicles. Poor approximation to the true value according to the observation effect of the state estimation system will lower the precision of

parameter utilized by dynamic control of vehicles and have great impact on the control effect. Therefore, one of the keys to realizing dynamic control effect of full vehicle lies in state estimation. Considering controllable single-wheel moment unique to distributed electric vehicles and features of accurate and available rotate velocity, multiple state variables shall be jointly observed in combination of information on electric driving wheel and in-vehicle sensor, so as to meet the demand for dynamic control of full vehicle. State parameters needing to be observed include motion state (including longitudinal vehicle velocity, side slip angle and yaw rate), acting force of vehicle (including lateral force and vertical force of tire), mass of full vehicle and road surface gradient. Accurate estimation of these state parameters will greatly improve the control effect of dynamic control system of full vehicle and the dynamic performance of vehicles. How to integrate multi-source information of distributed electric vehicles to jointly observe various states and comprehensively improve the observation precision of various state variables is one of the difficulties and key points of this dissertation.

(2) Design of vehicle desired dynamic target

In order to guarantee the power performance and stability of vehicles, reasonably desired dynamic target of vehicles shall be designed. Driver's expectation is also reflected by desired dynamic target. Failure to reflect driver's intention by the designed dynamic control target will dampen driving sensitivity and lower lateral stability of vehicles. In the process of developing desired dynamic target of full vehicle, longitudinal dynamic demand and lateral demand of full vehicle shall be determined on the basis of state parameter variables obtained by making use of state estimation system and driver's operation information. In the design of control target, impact of uncertain factors on control shall be taken into full consideration, so as to realize driver's expectation. In combined control over multiple targets, various control targets shall be comprehensively considered and the case that only single target is taken into consideration, which leads to occurrence of vehicle instability, shall be avoided. It is another key point and difficulty of this dissertation to determine reasonable demand target through comprehensive consideration of non-linear dynamic properties, road surface condition, dynamic response and steady-state response of distributed electric vehicles, in order to guarantee utmost safety of vehicles in various limited conditions.

(3) Driving force control allocation

Main purpose of driving force control allocation is to develop target driving moment of each wheel according to dynamic control target of full vehicle. Given the unique structural features of distributed electric vehicles, driving force can be allocated in a flexible way among multiple wheels during control allocation. In the meanwhile of meeting demand of full vehicle, optimized design can be made for some other targets. Various constraints imposed on distributed driving wheel need to be taken into consideration to ensure that all allocated force can be implemented. Constraints to be considered include failure constraint of driving wheel and driving anti-slip control constraint, etc. Due to constraints of driving wheel, driving force control allocation may be unable to meet the dynamic control target of full vehicle set by upper layer and in such a case multiple dynamic control targets shall be exposed to dynamic weight adjustment based on the current sate of vehicles and driver's operation to give top priority of main target. At the same time of optimization target design, minor optimization target shall be designed according to features of distributed electric driving wheel. Ultimately, the main target shall be integrated with the minor target and joint optimization shall be made for main target and minor target by virtue of non-linear optimization method. It is another key point and difficult problem to be solved in this dissertation, to comprehensively consider various constraints of driving force of distributed electric vehicles, and coordinate and optimize main target and minor target to maintain corresponding longitudinal driving force and direct yaw moment to the greatest extent in case of drive trackslip, drive failure and other constraints and guarantee drive capacity and safety of vehicles.

(4) Compensation control of motor property
Major purpose of compensation control of motor property is to control target driving moment made by driving motor following driving force control allocation layer, which is also the basic demand to realize dynamic control of distributed electric vehicles. The system needs to eliminate difference of actuator by rational control method to ensure target driving force to be implemented. In the design of controller, available features of moment of electric driving wheel shall be taken into consideration to design control method with superior robustness. In order to reduce or even eliminate the difference of motor moment, make moment of electric driving wheel stably follow target driving moment and ensure the realization of control intension of full vehicle, which is also one of the key points and difficulties of this dissertation, the designed compensation control method of motor property is required to overcome steady-state error and dynamic error of motor moment.

Chapter 3
State Estimation of Distributed Electric Vehicles

Abstract State parameter estimation is the foundation for dynamic control of vehicles. Rational utilization of structural features of distributed electric vehicles and observation of multiple state parameters by integration of multiple state parameters of vehicles are essential and basic work to realize dynamic control targets of distributed electric vehicles. Precision of state estimation directly determines dynamic control performance and the state parameter obtained from observation can not only be applied to coordinated control mentioned but also to other follow-up control processes. Current state estimation method of vehicles mainly originates from traditional vehicles, lacking consideration for inherent features of distributed electric vehicles. Aiming at the existing defects, this chapter made full use of the advantages in state estimation of distributed electric vehicles, and proposed the scheme of synthesizing multi-information and integrating multiple methods for combined observation of multiple states. Observation of state parameter lays foundation for dynamic control over distributed electric vehicles. Overall structure of state estimation system of distributed electric vehicles has been designed in Chap. 2, and this chapter will study state estimation methods of vehicles, including vehicle motion state observation (including longitudinal vehicle velocity, side slip angle and yaw rate), vehicle acting force observation (including lateral force and vertical force of tire), mass observation of full vehicle and gradient observation of road surface. According to state observation structure mentioned in Chap. 2 of this dissertation, estimation of mass of full vehicle, road surface gradient and vertical force of tire shall be completed first, on the basis of which non-linear vehicle dynamic model is built. Meanwhile, considering the strong non-linearity of the built vehicle dynamic model, unscented particle filter is designed to make combined observation of state of multiple wheels.

3.1 Mass Estimation of Full Vehicle Extraction Based on High Frequency Information

Mass of full vehicle (the mass mentioned hereunder refers to mass of full vehicle) impacts the design of vehicle control system. Accurate mass estimation will improve the control effect of such active-safety secondary controllers as ABS/ESP/ACC/

© Springer-Verlag Berlin Heidelberg 2016
W. Chu, *State Estimation and Coordinated Control for Distributed Electric Vehicles*, Springer Theses, DOI 10.1007/978-3-662-48708-2_3

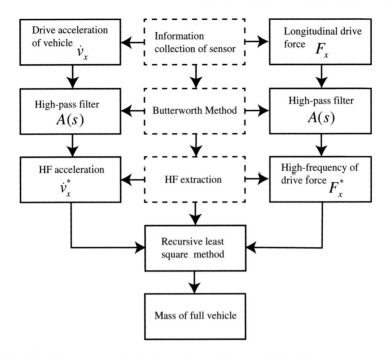

Fig. 3.1 Algorithm structure of estimation subsystem for mass of full vehicle

HAC [1]. However, common mass estimation calculation methods are coupled with road surface gradient, making precision of mass estimation limited to observation precision of road surface gradient and leading to poor estimation precision and slow convergence rate.

This dissertation puts forward a brand new mass estimation method by making use of structural features of Distributed Electric Vehicles. With improvement in precision and convergence rate, this method decouples mass and gradient during mass estimation. Figure 3.1 shows the algorithm structure of the designed mass estimation.

3.1.1 Analysis on Decoupling of Mass and Gradient

Longitudinal acceleration of vehicles is generated by combined action of driving force, rolling resistance, drag and gradient resistance, as shown in Fig. 3.2.

Longitudinal dynamic model of vehicles can be simplified as Eq. (3.1).

$$F_x = m\dot{v}_x + \frac{1}{2}\rho C_d A v_x^2 + mg(\sin\theta + f\cos\theta) \tag{3.1}$$

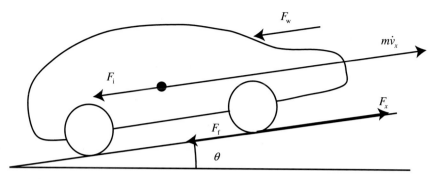

Fig. 3.2 Schematic diagram for longitudinal dynamic model

wherein, F_x refers to the longitudinal driving force, m refers to the mass of vehicle, v_x refers to the longitudinal velocity of vehicle, $\dot{v}_x$ refers to the driving acceleration, ρ refers to the air density, C_d refers to the drag coefficient, A refers to the windward area, g refers to the gravitational acceleration, θ refers to the road surface gradient and f refers to the rolling resistance coefficient of road surface. Perform differential to driving force F_x against driving acceleration $\dot{v}_x$ and then Eq. (3.2) can be obtained.

$$\frac{\mathrm{d}F_x}{\mathrm{d}\dot{v}_x} = m + \rho C_d A v_x \frac{\mathrm{d}v_x}{\mathrm{d}\dot{v}_x} + mg(\cos\theta - \sin\theta)\frac{\mathrm{d}\theta}{\mathrm{d}\dot{v}_x} \tag{3.2}$$

When performing gradient estimation by this method, assuming that gradient variation is commonly relatively small and has no direction correlation with driving acceleration $\dot{v}_x$ of vehicle, and variation of gradient is random [2, 3], then the differential of gradient against driving acceleration will be as shown in Eq. (3.3).

$$\frac{\mathrm{d}\theta}{\mathrm{d}\dot{v}_x} \doteq 0 \tag{3.3}$$

Variation frequency of driving moment of distributed electric vehicles is relatively higher when variation frequency of driving moment is higher, rate of variation of driving acceleration is far higher than that of velocity. It can be considered that signal differential of velocity is of tiny amount compared with that of driving acceleration, i.e.:

$$\frac{\mathrm{d}v_x}{\mathrm{d}\dot{v}_x} \doteq 0 \tag{3.4}$$

According to Eqs. (3.3) and (3.4), Eq. (3.2) can be simplified as Eq. (3.5) when the variation frequency of driving force is relatively higher.

$$m \doteq \frac{\mathrm{d}F_x}{\mathrm{d}\dot{v}_x} \tag{3.5}$$

Equation (3.5) shows that when the variation of driving force is relatively larger, impact of drag, rolling resistance and road surface gradient can be omitted. Mass of full vehicle at every moment during travelling can be acquired by Eq. (3.5), which approximates to differentiation of longitudinal driving force against driving acceleration.

From frequency analysis, differential of longitudinal driving force against driving acceleration approximates to the ratio of high-frequency information of the two. Therefore, perform high-pass filtering to them and then high-frequency information can be acquired followed by the desired differential.

The major requirement for high-pass filter by this method is to filter the impact of low-frequency part on drag, rolling resistance and road surface gradient. The impact of gradient variation frequency of road surface and other parameters shall be taken into consideration for the selection of cut-off frequency f_L. Low value of f_L makes removing impact of road surface gradient and drag impossible, while high one leads to filtering of too much information, so it is necessary to design cut-off frequency by taking different vehicle types into full comprehensive consideration.

Proper cut-off frequency f_L can be acquired by experimental method high-pass filter can be designed by Butterworth method [4]. Transfer function of high-pass filter is

$$A(s) = \frac{bs^2}{s^2 + a_1 s + a_0} \tag{3.6}$$

wherein, a_0, a_1, b are filtering parameters acquired by Butterworth method. It shall be noted that filters used for high-pass filtering of driving force and acceleration are both $A(s)$. By high-pass filtering to driving acceleration signal $\dot{v}_x$ and driving force signal F_x, high-frequency information $\dot{v}_x^*$ of driving acceleration and high-frequency information F_x^* of driving force are obtained. Since filtering is performed by the same filter, Eq. (3.7) can be obtained by high-pass filter principle.

$$\frac{\mathrm{d}F_x}{\mathrm{d}\dot{v}_x} \doteq \frac{F_x^*}{\dot{v}_x^*} \tag{3.7}$$

By combining Eqs. (3.5) and (3.7), fundamental Equation for mass estimation can be acquired, as shown in Eq. (3.8).

$$m \doteq \frac{F_x^*}{\dot{v}_x^*} \tag{3.8}$$

3.1.2 Mass Estimation by Recursive Least Square Method

Noise of actual signal is relatively higher in general. Mass estimation of full vehicle by Eq. (3.8) directly will be greatly impacted by noise of acceleration and driving force, which may deteriorate the observation result and lead to failure in direct application. Thus, the acquired result shall be exposed to retrogression treatment. Make $y = F_x^*$ and $\phi = \dot{v}_x^*$, Eq. (3.8) will convert to (3.9).

$$y = \phi\hat{m} \tag{3.9}$$

wherein, $\hat{m}$ is the estimated value of mass of full vehicle. Solution can be got by least square method [5]. In linear system, it is equivalent to find parameter $\hat{m}(k)$ and make function $V(\hat{m}(k), k)$ take the minimum value [6]. Wherein, k refers to the current sampling time.

$$V(\hat{m}(k), k) = \frac{1}{2}\sum_{i=1}^{k}(y(i) - \phi(i)\hat{m}(k))^2 \tag{3.10}$$

When Eq. (3.10) leads to the minimum value, there is Eq. (3.11).

$$\frac{\partial V(\hat{m}(k), k)}{\partial \hat{m}(k)} = 0 \tag{3.11}$$

Equation (3.11) for solving function $\hat{m}(k)$ can be obtained by Eq. (3.12).

$$\hat{m}(k) = (\sum_{i=1}^{k}\phi(i)^2)^{-1}(\sum_{i=1}^{k}\phi(i)y(i)) \tag{3.12}$$

Equation (3.12) shows that with the increase of k, calculated amount of $\hat{m}(k)$ also increases continuously. Given that mass estimation of full vehicles is performed in real time, recursive least square method (RLS) is adopted to estimate mass in actual application. In other words, estimated value of the former sampling time is corrected by utilizing the current sampling time. Algorithm of RLS is as shown in Eqs. (3.13)–(3.15).

$$\hat{m}(k) = \hat{m}(k-1) + L(k)(y(k) - \phi(k)\hat{m}(k-1)) \tag{3.13}$$

$$L(k) = P(k-1)\phi(k)(1 + \phi(k)P(k-1)\phi(k))^{-1} \tag{3.14}$$

$$P(k) = (1 - L(k)\phi(k))P(k-1) \tag{3.15}$$

In Eqs. (3.13)–(3.15), k refers to the current sampling time, while $k - 1$ refers to the former sampling time. Mass of full vehicle at various moment can be estimated via Eq. (3.13). By Eq. (3.14), the least square gain L is acquired. Equation (3.15) is the update to error covariance P.

3.2 Gradient Estimation Based on Multi-method Fusion

In the process of vehicle motion, acquisition of gradient information of road surface will improve the dynamic control effect of full vehicle and enhance the economical efficiency of full vehicle in the process of minor control [2]. Given that estimation of road surface gradient is usually coupled with that of mass of full vehicle, it is necessary to estimate mass during estimation of gradient in general. Decoupling analysis for mass of full vehicle and road surface gradient has been made in Sect. 3.1, and efficient estimation has been made for mass of full vehicle. Therefore, it is assumed that road surface gradient has been available when estimation of road surface gradient is performed. Figure 3.3 shows the algorithm structure of gradient estimation based on multi-method fusion.

3.2.1 Gradient Estimation Based on Dynamics

Dynamic model used for gradient estimation by dynamics is also Eq. (3.1). In Eq. (3.1), it can be considered that all values except gradient θ have been available or can be obtained. Make $y = F_x$, $u = m\dot{v}_x + \frac{1}{2}\rho C_d A v_x^2$, $b = mg(\sin\theta + f\cos\theta)$ and then Eq. (3.1) can be converted to Eq. (3.16).

$$y = u + b \tag{3.16}$$

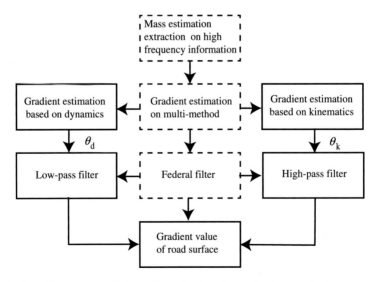

Fig. 3.3 Algorithm structure of observation subsystem for road surface gradient

wherein, y refers to the longitudinal driving force, which can be acquired accurately via feedback signal of driving motor; u refers to the functions of mass and velocity, which can be considered to have been known in case that mass is known; b refers to the function of mass and gradient; acquisition of b via estimation equals to acquisition of gradient value. In Eq. (3.16), estimation of b value can be performed by adopting least square method [5]. Given that road surface gradient is time-varying and b is the function of θ, then b is also time-varying. Therefore, least square method [7, 8] with forgetting factor is adopted to estimate b. In linear system, it is equivalent to acquisition of parameter $\hat{b}(k)$, making function $V(\hat{b}(k), k)$ take the minimum value [6]. Wherein, k refers to the current sampling time.

$$V(\hat{b}(k), k) = \frac{1}{2} \sum_{i=1}^{k} \lambda^{k-i} (y(i) - u(i) - \hat{b}(k)) \tag{3.17}$$

In Eq. (3.17), λ is forgetting factor. The larger the forgetting factor is, the higher the identification precision will be. However, it will lead to negative acceleration of the convergence velocity. Too small forgetting factor will result in lowered identification precision and recognition precision, while the convergence velocity will be improved. For balancing the contradiction between precision and convergence velocity, value of forgetting factor λ shall be considered comprehensively. When Eq. (3.17) takes the minimum value,

$$\frac{\partial V(\hat{b}(k), k)}{\partial \hat{b}(k)} = 0 \tag{3.18}$$

Equation (3.18) for solving $\hat{b}(k)$ is acquired from Eq. (3.19).

$$\hat{b}(k) = (\sum_{i=1}^{k} \lambda^{k-i})^{-1} (\sum_{i=1}^{k} \lambda^{k-i} (y(i) - u(i))) \tag{3.19}$$

Equation (3.19) shows that with the increase of k, calculated amount of $\hat{b}(k)$ also increases continuously. Given that gradient estimation of road is performed in real time, RLS with forgetting factor is adopted in actual application. Estimated value of the current sampling time is corrected by utilizing the former sampling time. Expression of RLS estimation method with forgetting factor is as shown from Eqs. (3.20) to (3.22).

$$\hat{b}(k) = \hat{b}(k-1) + L(k)(y(k) - u(k)) \tag{3.20}$$

$$L(k) = \frac{P(k-1)}{(\lambda + P(k-1))} \tag{3.21}$$

$$P(k) = \frac{1}{\lambda} (1 - L(k)) P(k-1) \tag{3.22}$$

From Eqs. (3.20) to (3.22), k refers to the current sampling time, while $k-1$ refers to the former sampling time. b can be obtained via Eq. (3.21). By Eq. (3.21), the least square gain L is acquired. Equation (3.22) is the renewal to error covariance P. After acquisition of b, the current road surface gradient can be obtained by utilizing the relational expression between b and road surface gradient, as shown in Eq. (3.23).

$$\theta_d = \arcsin \frac{D - f\sqrt{1 - D^2 + f^2}}{1 + f^2} \tag{3.23}$$

wherein,

$$D = \frac{b}{mg} \tag{3.24}$$

Estimated value θ_d of road surface gradient based on dynamics can be acquired via Eqs. (3.23) and (3.24).

3.2.2 Gradient Estimation Based on Kinematics

Acceleration sensor is fixed on the vehicle body, measured value of which is not only impacted by driving acceleration but also by road surface gradient. Figure 3.4 is the schematic diagram for road surface gradient, wherein $a_{x,m}$ refers to the vehicle longitudinal acceleration measured by acceleration sensor and $\dot{v}_x$ is the vehicle driving acceleration.

Relation between $a_{x,m}$ and $\dot{v}_x$ is as shown in Eq. (3.25).

$$a_{x,m} = \dot{v}_x + g\sin\theta \tag{3.25}$$

Kinematic estimation method for gradient of road can be obtained via Eq. (3.25), as shown in (3.26).

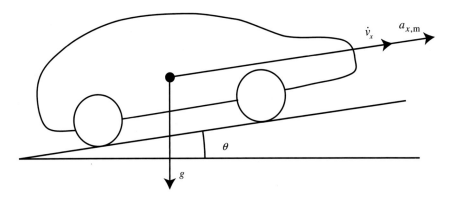

Fig. 3.4 Schematic diagram for road surface gradient

$$\theta_k = \arcsin(\frac{a_{x,m} - \dot{v}_x}{g})$$ (3.26)

Estimated value θ_k of road surface gradient based on kinematics can be acquired by Eq. (3.26).

3.2.3 Fusion of Kinematics and Dynamics

Road surface gradient can be considered to be composed of high-frequency signal and low-frequency signal in the travelling process of vehicle. Gradient estimation precision based on dynamics relies on vehicle model, while various parameters in vehicle model are greatly impacted by high-frequency noise. Therefore, high-frequency part of θ_d shall be removed by low-pass filtering with the low-frequency part being reserved. Defect of gradient estimation based on kinematics is that the measured value $a_{x,m}$ of acceleration sensor is greatly affected by static deviation, while static deviation of inertial sensor belongs to low-frequency noise. In order to maintain the accuracy of estimation result of road surface gradient, high-pass filtering is adopted for θ_k to remove the low-frequency part and reserve the high-frequency.

Gradient estimation via fusion of kinematics and dynamics can be realized by Eq. (3.27), design idea of which is from the Ref. [9].

$$\theta = \frac{1}{\tau s + 1}\theta_d + \frac{\tau s}{\tau s + 1}\theta_k$$ (3.27)

wherein, τ refers to the time constant. When the variation frequency of road surface gradient is relatively low, effect of dynamic estimation method is obvious, which can estimates the static gradient value in a relatively accurate way. However, when the variation frequency of road surface gradient is relatively high, effect of kinematic estimation method is more obvious, which can follow the transient-state variation of gradient.

In this dissertation, dynamics and kinematics are adopted to make real-time estimation of road surface gradient respectively, then by fusion of the estimation results of which relatively accurate gradient estimated value can be acquired. Compared with existing estimation method, not only impact of high-frequency noise information in dynamics on gradient estimation precision of road surface is eliminated but also error brought by such low-frequency signals as static deviation of acceleration sensor in kinematics on road surface gradient is also avoided.

3.3 Vertical Force Estimation of Tire Based on Multi-information Infusion

Vertical force of wheel is mainly impacted by static load and dynamic load. Wherein, static load mainly lies in mass of full vehicle and road surface gradient, while dynamic load mainly depends on motion of vehicles, including longitudinal accel-

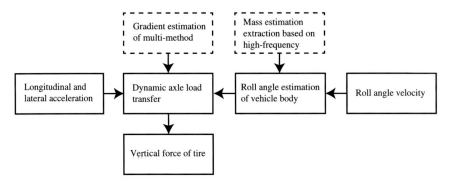

Fig. 3.5 Algorithm structure of vertical force estimation subsystem of tire

erated motion, lateral accelerated motion and rolling motion of vehicles. In previous estimation of vertical force, dynamic axle load transfer is estimated only by utilizing the information of longitudinal acceleration and lateral acceleration [10]. Such an estimation method is absent from consideration for rolling motion of vehicles and usually leads to inaccurate result. Roll angle and velocity of roll angle of vehicle body are estimated in this dissertation before vertical force was estimated, and basing on this vertical force estimation was completed. Figure 3.5 shows the algorithm structure to vertical force estimation of tire based on multi-information infusion.

3.3.1 Estimation of Roll Angle and Roll Rate of Vehicle Body

For estimation of roll angle and roll angle velocity of vehicle body, partial basic assumption is needed, including that the roll angle of vehicle body is very small, $\sin \phi \doteq \phi$ and $\cos \phi \doteq 1$; side slope angle of road surface is 0 and that pitch angle of vehicle is 0.

Schematic diagram for rolling motion of vehicle body is shown as Fig. 3.6. Rolling motion equation of vehicle body is (3.28).

$$I_z \ddot{\phi} + C_{\text{roll}} \dot{\phi} + K_{\text{roll}} \phi = m h_s a_{y,\text{m}} \tag{3.28}$$

By making use of rolling angle velocity sensor, rolling velocity of vehicle body can be observed in real time, i.e.

$$\dot{\phi} = \dot{\phi}_\text{m} \tag{3.29}$$

State observer can be designed to observe roll angle ϕ of vehicle body and roll angle velocity $\dot{\phi}$ according to Eqs. (3.28) and (3.29), as shown in Eq. (3.30).

$$\begin{cases} \dot{x} = Ax + Bu + q \\ y = Cx + r \end{cases} \tag{3.30}$$

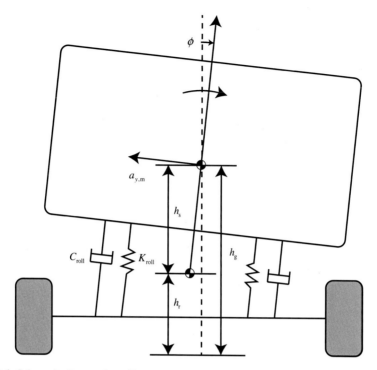

Fig. 3.6 Schematic diagram for rolling motion of vehicle body

$$\boldsymbol{x} = \begin{bmatrix} \dot{\phi} \\ \phi \end{bmatrix}, \boldsymbol{A} = \begin{bmatrix} -\frac{C_{\text{roll}}}{I_x} & -\frac{K_{\text{roll}}}{I_x} \\ 1 & 0 \end{bmatrix}, \boldsymbol{B} = \frac{mh_{\text{r}}}{I_x}, \boldsymbol{u} = a_{y,\text{m}}, \boldsymbol{C} = [1, 0], \boldsymbol{y} = \dot{\phi}_{\text{m}}$$

wherein, $\boldsymbol{x}$ refers to the state function, $\boldsymbol{q}$ refers to the process noise and $\boldsymbol{r}$ refers to the measurement noise. Covariance matrix of process noise is $\boldsymbol{Q}$ and measurement noise is $\boldsymbol{R}$. Perform filtering estimation for state equation (3.30) by Kalman Filtering [11, 12], by which roll angle ϕ and roll angle velocity $\dot{\phi}$ of vehicle body can be obtained and the computational process for it is as follows.

(1) Discretization

$$\boldsymbol{x}_{k+1} = \underbrace{(\boldsymbol{I} + \boldsymbol{A}_k \Delta T)}_{A_k^*} \boldsymbol{x}_k + \underbrace{\boldsymbol{B}_k \Delta T}_{B_k^*} \boldsymbol{u}_k + \boldsymbol{q}_k \tag{3.31}$$

$$\boldsymbol{y}_k = \boldsymbol{C}\boldsymbol{x}_k + \boldsymbol{r}_k \tag{3.32}$$

wherein, ΔT refers to the sampling time.

(2) Initialization ($k = 0$)
Set initial state as x_0 and initial covariance as P_0.

(3) Recursive estimation $k = 1, 2, \ldots$
Time update equation is

$$x_{k|k-1} = A^*_{k-1}x_{k-1} + B^*_{k-1}u_{k-1} \tag{3.33}$$

$$P_{k|k-1} = A^*_{k-1}x_{k-1}A^{*\mathrm{T}}_{k-1} + Q_{k-1} \tag{3.34}$$

State update equation is

$$K_k = P_{k|k-1}C^{\mathrm{T}}_k(C_k P_{k|k-1}C^{\mathrm{T}}_k + R_k)^{-1} \tag{3.35}$$

$$P_k = (I - K_k C^{\mathrm{T}}_k)P_{k|k-1} \tag{3.36}$$

$$x_k = x_{k|k-1} + K_k(y_k - C_k x_{k|k-1}) \tag{3.37}$$

From Eqs. (3.31) to (3.37), a complete set of Kalman filtering observer is composed, and state vector x can be observed from it. Substitute the observed roll angle ϕ of vehicle body and roll angle velocity $\dot{\phi}$ into Eq. (3.41) and then axle load transfer caused by rolling of vehicle body can be obtained.

3.3.2 Analysis on Dynamic Axle Load Transfer

In addition to vertical load caused by mass of full vehicle perpendicular to direction of road surface, vertical force of tire is also affected by four kinds of axle load transfer, including axle load transfer $F_{z,\theta}$ caused by road surface gradient, axle load transfer F_{z,a_x} caused by longitudinal acceleration, axle load transfer F_{z,a_y} caused by lateral acceleration and axle load transfer caused by rolling motion $F_{z,\phi}$. Axle load transfer caused by road surface gradient is as shown in Eq. (3.38).

$$F_{z,\theta} = mg\frac{h_g}{l}\sin\theta \tag{3.38}$$

Axle load transfer caused by longitudinal acceleration is as shown in Eq. (3.39).

$$F_{z,a_x} = ma_x\frac{h_g}{l} \tag{3.39}$$

Axle load transfer caused by lateral acceleration is as shown in Eq. (3.40).

$$F_{z,a_y} = ma_y \frac{h_r}{b} \tag{3.40}$$

As shown in Fig. 3.6, axle load transfer caused by rolling of vehicle body is Eq. (3.41).

$$F_{z,\phi} = \frac{K_{roll}\phi + C_{roll}\dot{\phi}}{b} \tag{3.41}$$

wherein, K_{roll} refers to the roll angle stiffness of vehicle body; C_{roll} refers to the roll damping of vehicle body. Roll angle ϕ and roll angle velocity $\dot{\phi}$ of vehicle body have been obtained from estimation in Sect. 3.3.1. Mass of full vehicle m and gradient θ of road surface have been estimated in Sects. 3.1 and 3.2 respectively, and longitudinal acceleration a_x and lateral acceleration a_y can be obtained by making use of the existing inertial sensor. Meanwhile, axle load transfer $F_{z,\phi}$ caused by rolling motion of vehicle body has been obtained by estimation in this section. Accordingly, vertical force of tire can be estimated by Eq. (3.42).

$$\begin{bmatrix} F_{1,z} \\ F_{2,z} \\ F_{3,z} \\ F_{4,z} \end{bmatrix} = \begin{bmatrix} l_r \\ l_r \\ l_f \\ l_f \end{bmatrix} \frac{mg\cos\theta}{2l} + \begin{bmatrix} -1 \\ -1 \\ 1 \\ 1 \end{bmatrix} \frac{F_{z,\theta}}{2} + \begin{bmatrix} -1 \\ -1 \\ 1 \\ 1 \end{bmatrix} \frac{F_{z,a_x}}{2} + \begin{bmatrix} -l_r \\ l_r \\ -l_f \\ l_f \end{bmatrix} \frac{F_{z,a_y}}{l} + \begin{bmatrix} -1 \\ 1 \\ -1 \\ 1 \end{bmatrix} \frac{F_{z,\phi}}{2} \tag{3.42}$$

3.4 Estimation of Vehicle Motion State and Lateral Force Based on UPF

Observation methods for state parameter generally fall into kinematic one and dynamic one. Kinematic observation method focuses on motion state itself and makes estimation on motion state of vehicles by direct application of INS information; the method features sound robustness and has little dependence on vehicle model but with relies on the precision of INS [13]. However, dynamic observation method pays more attention to acting force causing vehicle motion, requires relatively accurate information of acting force, relies on model precision but has lowered reliance on INS. In distributed electric vehicles, moment and rotate velocity of driving wheel are accurate and available and sensing range of information sees large expansion compared with that of traditional vehicles. By setting up relatively appropriate vehicle model and reasonable fusion of kinematic and dynamics, observation precision can be efficiently improved [14]. For combined estimation on longitudinal velocity, lateral velocity, yaw rate and lateral force of tire in running process of vehicle, kinematics and dynamics were adopted in this section to set up non-linear-state observation model.

In setting up of model, the previously observed mass m of full vehicle, gradient θ of road surface and vertical force $F_{i,z}$ of tire are taken as known value. By setting up of non-linear-state observation model, rational state equation can be designed. State equation is set up by non-linear-state observation model, which includes three parts, i.e. "non-linear vehicle dynamic model", "Magic Formula tire model" and "dynamic tire model". "Non-linear vehicle dynamic model" pays close attention to dynamic motion of entire vehicle; by "Magic Formula tire model" quasi-static lateral force of tire can be acquired, while "dynamic tire model" expands the applicable range of lateral force from quasi-static state to dynamic process, following the variation of lateral force in a satisfactory way.

Measurements of distributed electric vehicles includes longitudinal and lateral acceleration, yaw rate and wheel velocity of each wheel, in which wheel velocity of each wheel can be acquired via feedback of distributed drive motor. Although the information of longitudinal and lateral acceleration and yaw rate can be obtained by INS directly, static deviation has relatively great impact on INS, thus requiring calibration for INS to remove deviation.

After state observation equation and measured value of measurement vector are obtained, unscented particle filtering observer is designed in consideration of strong non-linear property of the constructed vehicle model to perform combined observation for multiple state variables of vehicles. Moreover, to further improve the working condition adaptability of state observer, self-adaptive control of measurement noise is performed, improving precision of state observation.

In this dissertation, unscented particle filter (UPF) was designed firstly, followed by setting up of non-linear-state observation model, calibration of inertial sensor and self-adaptive design of measurement noise. Vehicle motion state based on unscented particle filtering and estimation flow of lateral force is shown in Fig. 3.7.

3.4.1 Design of Unscented Particle Filter

Non-linear-state observation model is designed according to non-linear vehicle dynamic model shown in Fig. 3.7.

Given the features of distributed electric vehicles, quantity that is related to state estimation and can be directly measured include steering wheel angle, longitudinal and lateral velocity, yaw rate, motor driving moment/braking moment and rotate velocity of motor.

Input vector is designed firstly, and the longitudinal driving force generated by distributed electric driving wheel can be calculated by Eq. (3.43).

$$F_{i,x} = \frac{T_i - I_w \dot{\omega}_i}{R} \tag{3.43}$$

wherein, I_w refers to the rotational inertia of driving wheel (including motor) and $\dot{\omega}_i$ refers to the angle acceleration of the wheel. Since moment and rotate velocity of

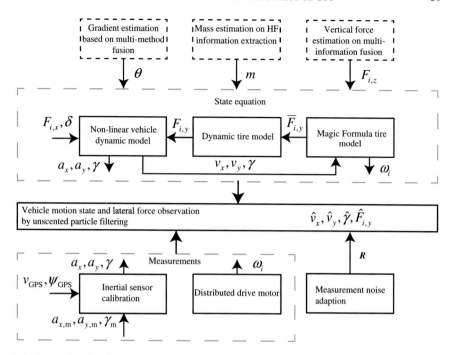

Fig. 3.7 Estimation flow motion state and lateral force based on unscented particle filtering

driving motor are accurate and known, in-wheel driving force $F_{i,x}$ can be considered to be known.

For front-wheel steering vehicles, steering angle of each wheel δ_i can be simplified, as shown in Eq. (3.44).

$$\begin{cases} \delta_1 = \delta \\ \delta_2 = \delta \\ \delta_3 = 0 \\ \delta_4 = 0 \end{cases} \tag{3.44}$$

As a result, the input vector $\boldsymbol{u}$ is defined to be composed of steering angle of front wheel and driving force of each wheel.

$$\boldsymbol{u} = [\delta \quad F_{1,x} \quad F_{2,x} \quad F_{3,x} \quad F_{4,x}]^{\mathrm{T}} \tag{3.45}$$

States to be observed includes longitudinal and lateral acceleration, yaw rate and driving force of each wheel. State vector $\boldsymbol{x}$ is defined as

$$\boldsymbol{x} = [v_x \quad v_y \quad \gamma \quad F_{1,y} \quad F_{2,y} \quad F_{3,y} \quad F_{4,y}]^{\mathrm{T}} \tag{3.46}$$

Measurement vector is composed of measured value of inertial sensor and rotate velocity of each wheel. Measurement vector y is defined as

$$y = [a_x \quad a_y \quad \gamma \quad \omega_1 \quad \omega_2 \quad \omega_3 \quad \omega_4]^{\mathrm{T}} \tag{3.47}$$

Recurrence equation is to predict state vector of the next moment via the current state vector and input vector. State recurrence equation can be obtained based on non-linear-state observation model, as follows

$$\dot{x}(t) = \widetilde{f}(x(t), u(t)) \tag{3.48}$$

Measurement equation is to calculate the measurement vector of the current moment via the current state vector and input vector. Measurement equation can be acquired based on non-linear-state observation model, as follows

$$y(t) = \widetilde{h}(x(t), u(t)) \tag{3.49}$$

Meanwhile, wheel velocity in measurement vector can be obtained directly from distributed drive motor, while inertial sensor part can be acquired directly via measured value of calibrated inertial sensor. As a result, complete state equations are set up, including Eqs. (3.48) and (3.49). On the basis of setting up complete state observation model, select proper filter to perform State Estimation. As one of the most efficient mathematical tools in state observation field [11, 12, 15], Kalman Filtering (KF) has been widely applied in field of vehicle state estimation [16], but limited to linear observation.

Extended Kalman Filtering (EKF) is constructed based on classic linear Kalman filtering [11, 15, 17], basic concept of which is to carry out first-order Taylor expansion for inertial model against estimated state value and then perform state observation by application of classic linear KF Equation. However, EKF must meet the linear hypothesis. In other words, difference between linear equation and original linear equation after Taylor expansion is the minimal value. Meanwhile, EKF must perform calculation of Jacobian matrix.

In order to overcome the defects of EKF, Julier et al. developed Unscented Kalman Filtering (UKF) in the 1990s [18–20]. Different from EKF, UKF performs state estimation by direct use of non-linear equation, shunning from the impact of linearization error and without the need to calculate Jacobian matrix. Basic assumption of EKF and UKF is that state noise belongs to Gaussian distribution. However, the real working condition is quite complex, so Gaussian assumption tends to be incorrect in practical application.

To abolish Gaussian assumption, some experts proposed particle filter (PF) in recent years [21, 22], which was set up on the basis of importance sampling and Monte Carlo assumption and abolished Gaussian assumption, providing efficient way for high-order non-linear state observation. Major defect of PF lies in particle impoverishment; in other words, with the increase of filtering iterations, majority of particle offspring with smaller weights becomes little or disappears and only small amount of particle offspring with larger weights is abundant and diversity of particle matters weakens, thus it is insufficient to be used for approximate representation of posterior density [23, 24]. For solving particle impoverishment incurred by PF, Van der Merwe et al. from University of Cambridge put forward unscented particle filter (UPF) [24]. This method designs particle distribution by taking advantage of unscented transformation (UT), thus solving particle impoverishment caused by PF and retaining the advantages of PF. Compared with EKF and UKF, UPF is more suitable for observation of strong non-linear problems, with the following advantages: no need of calculation of Jacobian matrix, retaining high-order information of non-linear problems in estimated result and abolishing Gaussian assumption of state noise.

Compared with EKF and UKF, major disadvantage of UPF lies in larger calculation amount. However, owing to superior approximation to non-linear features of vehicles and tires by UPF, relatively larger simulation step size can be selected during state observation, making for the defect of large calculation amount to some extent. Effective observation can be made for strong non-linear property of the constructed vehicle model by adopting UPF. State Estimation for Distributed Electric Vehicles by making use of UPF is divided into the following steps and derivation process is made by reference to reference [18, 19, 24].

(1) State equation discretization

Discretize state equation (3.48) and (3.49) and then discretization state equation is acquired for state observation, as shown in Eqs. (3.50) and (3.51).

$$x_{k+1} = f(x_k, u_k) + q_k \tag{3.50}$$

$$y_k = h(x_k, u_k) + r_k \tag{3.51}$$

In Eqs. (3.50) and (3.51), $x(k) \in \mathbb{R}^{n_x}$ refers to the state vector at moment k ; $u(k) \in \mathbb{R}^{n_u}$ refers to the input vector; $y(k) \in \mathbb{R}^{n_y}$ refers to the output vector; $q(k) \in \mathbb{R}^{n_x}$ refers to the process noise; $r(k) \in \mathbb{R}^{n_y}$ refers to the measurement noise. Wherein, auto-covariance and cross covariance of process noise $q(k)$ and measurement noise $r(k)$ is

$$E[q_i q_j^T] = \delta_{ij} Q, \ \forall \, i, j \tag{3.52}$$

$$E[r_i r_j^T] = \delta_{ij} R, \ \forall \, i, j \tag{3.53}$$

$$E[q_i r_j^T] = 0, \ \forall \, i, j \tag{3.54}$$

wherein, $Q \in \mathbb{R}^{n_x \times n_x}$ and $R \in \mathbb{R}^{n_y \times n_y}$ refers to the symmetric positive matrix; δ_{ij} is Kronecker δ function.

$$\delta_{ij} = \begin{cases} 1 & i = j \\ 0 & i \neq j \end{cases} \tag{3.55}$$

(2) Initialized filter (when $k = 0$)

Extract N particles $\{x_0^{(i)}\}_{i=1}^N$ from priori distribution $p(x_0)$. The more the number of particles is, the closer the state distribution generated by filter will be to the posteriori distribution of state [25]. Initial value $x_0^{(i)}$ and initial covariance $P_0^{(i)}$ of state vector are obtained according to the extracted particle $\{x_0^{(i)}\}_{i=1}^N$.

$$\bar{x}_0^{(i)} = E[x_0^{(i)}] \tag{3.56}$$

$$P_0^{(i)} = E[(x_0^{(i)} \quad \bar{x}_0^{(i)})(x_0^{(i)} - \bar{r}_0^{(i)})^T] \tag{3.57}$$

In consideration of the process noise and measurement noise, the extended initial state vector and covariance is

$$\bar{x}_0^{(i)a} = E[x_0^{(i)a}] = [(\bar{x}_0^{(i)})^T \quad 0 \quad 0] \tag{3.58}$$

$$P_0^{(i)a} = E[(x_0^{(i)a} - \bar{x}_0^{(i)a})(x_0^{(i)a} - \bar{x}_0^{(i)a})^T] = \begin{bmatrix} P_0^{(i)} & 0 & 0 \\ 0 & Q & 0 \\ 0 & 0 & R \end{bmatrix} \tag{3.59}$$

(3) Calculation of recurrence filter (when $k = 1, 2, \ldots$)

(a) Sigma point sampling

Basic concept of unscented transformation is to select $2n_a + 1$ weighting sampling points (i.e. Sigma point) to approximate the distribution of random variable $\bar{x}_{k-1}^{(i)a}$ by making use of the mean value $\bar{x}_{k-1}^{(i)a}$ and variance $P_{k-1}^{(i)a}$ of state vector. Symmetric sampling strategy is adopted in this dissertation to select Sigma point [26] and unscented transformation is adopted to update Sigma point of N particles. For $i = 1, \ldots, N$ there is

$$\mathcal{X}_{k-1}^{(i)a} = [\bar{x}_{k-1}^{(i)a}, \bar{x}_{k-1}^{(i)a} \pm (\sqrt{(n_a + \lambda)P_{k-1}^{(i)a}})] \tag{3.60}$$

(b) Time update process

Select proper variables of κ, α, β. κ is used to change $W_0^{(c)}$. $\kappa \geq 0$ is generally selected to guarantee the positive semi-definiteness of covariance matrix. α is used to adjust distribution distance of particles, reduce the impact of higher moment and lower predictive error and a relatively small positive value is generally selected, i.e. $10^{-4} \leq \alpha \leq 1$. β includes the information of higher moment of $\bar{x}^{(i)a}$ priori distribution. In conclusion, the selection result shall be $\kappa = 0$, $\alpha = 0.001$, $\beta = 2$. Wherein, λ is amplification coefficient, by which weight coefficient of each particle can be adjusted.

$$\lambda = \alpha^2(n_a + \kappa) - n_a \tag{3.61}$$

Dimension of state vector x is n_x, that of measurement noise is n_v and that of the extended state vector x^a is n_a.

$$n_a = 2n_x + n_v \tag{3.62}$$

Weight of Sigma point is

$$W_0^{(m)} = \frac{\lambda}{n_a + \lambda} \tag{3.63}$$

$$W_0^{(c)} = \frac{\lambda}{n_a + \lambda} + (1 - \alpha^2 + \beta) \tag{3.64}$$

$$W_i^{(m)} = W_i^{(c)} = \frac{1}{2(n_a + \lambda)}, i = 1, \ldots, 2n_a \tag{3.65}$$

Update all Sigma points by non-linear-state equation (3.50).

$$\mathscr{X}_{k|k-1}^{(i)x} = f(\mathscr{X}_{k-1}^{(i)x}, u(k-1)) + \mathscr{X}_{k-1}^{(i)w} \tag{3.66}$$

Predicted value of state vector by weighting calculation is

$$\bar{x}_{k|k-1}^{(i)} = \sum_{j=0}^{2n_a} W_j^{(m)} \mathscr{X}_{j,k|k-1}^{(i)x} \tag{3.67}$$

Perform non-linear transformation by non-linear-state equation (3.51):

$$\mathscr{Y}_{k|k-1}^{(i)x} = h(\mathscr{X}_{k-1}^{(i)x}, u(k-1)) + \mathscr{X}_{k-1}^{(i)v} \tag{3.68}$$

Predicted system value by weighting calculation is

$$\bar{y}_{k|k-1}^{(i)} = \sum_{j=0}^{2n_a} W_j^{(m)} \mathscr{X}_{j,k|k-1}^{(i)x} \tag{3.69}$$

Predicted value of covariance matrix by weighting calculation is

$$P_{k|k-1}^{(i)} = \sum_{j=0}^{2n_a} W_j^{(c)} (\mathscr{X}_{j,k|k-1}^{(i)x} - \bar{x}_{k|k-1}^{(i)})(\mathscr{X}_{j,k|k-1}^{(i)x} - \bar{x}_{k|k-1}^{(i)})^{\mathrm{T}} \tag{3.70}$$

(c) Measurement update process
Covariance matrix of priori estimated error is

$$P_{x_k, y_k} = \sum_{j=0}^{2n_a} W_j^{(c)} (\mathscr{X}_{j,k|k-1}^{(i)} - \bar{x}_{k|k-1}^{(i)})(\mathscr{Y}_{j,k|k-1}^{(i)} - \bar{y}_{k|k-1}^{(i)})^{\mathrm{T}} \tag{3.71}$$

Covariance matrix of posteriori estimated error is

$$P_{\tilde{y}_k,\tilde{y}_k} = \sum_{j=0}^{2n_a} W_j^{(c)} (\mathscr{Y}_{j,k|k-1}^{(i)} - \bar{y}_{k|k-1}^{(i)})(\mathscr{Y}_{j,k|k-1}^{(i)} - \bar{y}_{k|k-1}^{(i)})^{\mathrm{T}} \tag{3.72}$$

Filter gain matrix is

$$K_k = P_{x_k,y_k} P_{\tilde{y}_k,\tilde{y}_k}^{-1} \tag{3.73}$$

Filter value after state update is

$$\bar{x}_k^{(i)} = \bar{x}_{k|k-1}^{(i)} + K_k(y_k - \bar{y}_{k|k-1}^{(i)}) \tag{3.74}$$

Posteriori variance matrix after state update is

$$\hat{P}_k^{(i)} = P_{k|k-1}^{(i)} - K_k P_{\tilde{y}_k,\tilde{y}_k} K_k^{\mathrm{T}} \tag{3.75}$$

(d) Importance sampling
Select state transition probability density as the probability density of importance sampling, i.e. extract $\hat{x}_k^{(i)}$ for N particles separately, wherein $i = 1, \ldots, N$.

$$\hat{x}_k^{(i)} \sim q(x_k^{(i)}|x_{0:k-1}^{(i)}, y_{1:k}) = \mathscr{N}(\bar{x}_k^{(i)}, \hat{P}_k^{(i)}) \tag{3.76}$$

Define $\hat{x}_{0:k}^{(i)}$ and $\hat{P}_{0:k}^{(i)}$ as

$$\hat{x}_{0:k}^{(i)} \triangleq (x_{0:k-1}^{(i)}, \hat{x}_k^{(i)}) \tag{3.77}$$

$$\hat{P}_{0:k}^{(i)} \triangleq (P_{0:k-1}^{(i)}, \hat{P}_k^{(i)}) \tag{3.78}$$

For $i = 1, \ldots, N$, weight of new samples of each particle is calculated as

$$w_k^{(i)} \propto \frac{p(y_k|\hat{x}_k^{(i)})p(\hat{x}_k^{(i)}|x_{k-1}^{(i)})}{q(\hat{x}_k^{(i)}|x_{0:k-1}^{(i)}, y_{1:t})} \tag{3.79}$$

For $i = 1, \ldots, N$, normalize weight of each particle by

$$\tilde{w}_k^{(i)} = w_k^{(i)} \left[\sum_{j=1}^{N} w_k^{(j)} \right]^{-1} \tag{3.80}$$

(e) Re-sampling
Particle degeneracy means that weight of almost all particles approaches to 0 after several steps of recurrence except individual particles, which is a major defect of

particle filter [21, 27]. To avoid that too much operation concentrates on particles with very small weight, re-sampling shall be introduced to reduce particles with small weight and increase particles with large weight. Firstly, duplicate the corresponding particle swarm $\{\hat{x}_{0:k}^{(i)}, \hat{P}_{0:k}^{(i)}\}_{i=1}^{N}$ according to the size of particle weight $\tilde{w}_k^{(i)}$, increase particle swarms with large weight and reduce particle swarms with small weight. Secondly, acquire N random particle swarms $\{\tilde{x}_{0:k}^{(i)}, \tilde{P}_{0:k}^{(i)}\}_{i=1}^{N}$ separately according to posterior probability distribution $p(x_{0:k}^{(i)}|y_{1:k})$ and substitute the previous particle swarm by these new ones. Reset corresponding weight as $w_k^{(i)} = \tilde{w}_k^{(i)} = \frac{1}{N}$ for these new random particle swarms.

(f) State output
Output of UPF is one set of sampling points, which can be used for approximating the real posterior probability distribution. Estimated value of state vector of particle weight is to be

$$\bar{x}_k^{(i)} = \sum_{i=1}^{N} w_k^{(i)} x_k^{(i)} \tag{3.81}$$

Covariance matrix of state vector with particle weight considered is

$$\bar{P}_k^{(i)} = \sum_{i=1}^{N} w_k^{(i)} (x_k^{(i)} - \bar{x}_k^{(i)})(x_k^{(i)} - \bar{x}_k^{(i)})^{\mathrm{T}} \tag{3.82}$$

Repeat Process (a)–(f) during recurrence calculation and then estimated value $\hat{x}_k$ of state vector at k moment can be acquired, as shown in Eq. (3.83).

$$\hat{x}_k = \bar{x}_k^{(i)} \tag{3.83}$$

3.4.2 Analysis on Non-linear Vehicle Dynamic Model

The basic assumptions necessary for establishing vehicle model include that dynamic condition of suspension can be neglected; vehicle body is one rigid one; resultant force of windward resistance and rolling resistance acts on center of mass, center of mass of vehicle is fixed on a known position of the vehicle body; only longitudinal, lateral and lateral motion are taken into consideration.

Coordinate system of distributed electric vehicles designed is as shown in Fig. 3.8. Take center of mass (CoG) of distributed electric vehicles as origin of the coordinate, create coordinate system $x_V y_V$ concreted on vehicle body and adopt ISO definition for the coordinate system [28, 29]. Wherein, x_V axis points to the heading direction of vehicle, while y_V axis points to the left side of the heading direction of vehicle.

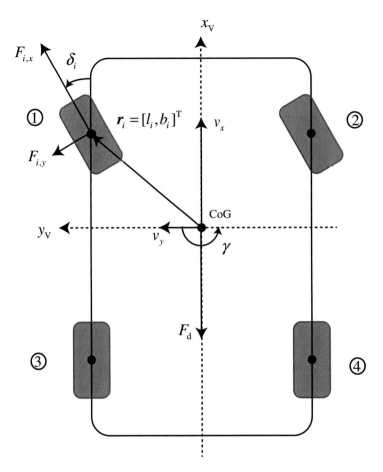

Fig. 3.8 Coordinate system of state observation

As shown in Figs. 3.9 and 3.10, define tire coordinate system $x_W\,y_W$ concreted on each tire with its origin in the center of each tire. Under coordinate system of vehicle body, coordinate of origin of each tire coordinate system is $r_i = [l_i, b_i]^T i \in \{1, 2, 3, 4\}$. Wherein, i represents corresponding tire number in Fig. 3.8; steering angle of coordinate system of each tire against vehicle body is δ_i. $x_{i,W}$ axis points to rolling direction of tire and $y_{i,W}$ axis to left side of rolling direction of tire. Meanwhile, we define origin to concrete on coordinate system $x_R\,y_R$ of each wheel, with $x_{i,R}$ axis pointing to heading direction of vehicle and $y_{i,R}$ axis to left side of heading direction of vehicle.

(1) Analysis on tire dynamic model
As shown in Fig. 3.9, driving force generated by electric driving wheel is $F_{i,x}$; lateral force of tire is $F_{i,y}$; resultant force generated under tire coordinate system is $F_i = [F_{i,x}, F_{i,y}]^T$. Transformational matrix from tire coordinate sys-

Fig. 3.9 Tire dynamic model

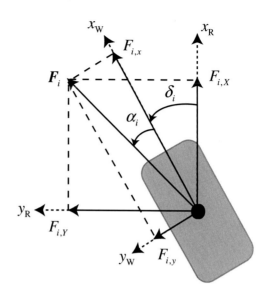

Fig. 3.10 Tire kinematic model

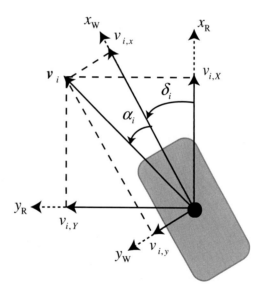

tem $x_W y_W$ to $x_R y_R$ is $\boldsymbol{R}(\delta_i)$, which depends on steering angle of the tire. Therefore, force vector $\boldsymbol{F}_{i,R} = [F_{i,X}, F_{i,Y}]^T$ by projection of $\boldsymbol{F}_i$ on coordinate system $x_R y_R$ can be calculated by Eq. (3.84).

$$\boldsymbol{F}_{i,R} = \boldsymbol{R}(\delta_i)\boldsymbol{F}_i \tag{3.84}$$

wherein, $R(\delta_i)$ refers to the function of steering angle δ_i of tire.

$$R(\delta_i) = \begin{bmatrix} \cos\delta_i & -\sin\delta_i \\ \sin\delta_i & \cos\delta_i \end{bmatrix} \tag{3.85}$$

(2) Analysis on dynamic model of vehicle

In addition to acting force of tire, forces acting on vehicle also include drag, rolling resistance and gradient resistance. Given that acting force of drag and rolling resistance has been assumed to be in center of mass, the resultant force of drag, rolling resistance and gradient reslstance is denoted as F_d. In vehicle coordinate system, the resultant force of drag, rolling resistance and gradient resistance is

$$F_d = \begin{bmatrix} \frac{1}{2}\rho C_d A v_x^2 + mgf\cos\theta + mg\sin\theta \\ 0 \end{bmatrix} \tag{3.86}$$

Simplify force system $\{F_{1,R}, F_{2,R}, F_{3,R}, F_{4,R}, F_d\}$ in vehicle coordinate system $x_V y_V$ towards center of mass of vehicle based on shifting theorem of plane general force system [30]. Principal vector of this force system R is the algebraic sum of each vector and principal moment M_o is the sum of couples of each force vector against center of mass.

$$R = [F_x, F_y]^T \tag{3.87}$$

$$M_o = M \tag{3.88}$$

wherein, F_x, F_y and M refer to the longitudinal resultant force, lateral resultant force and resultant yaw moment acting on center of mass respectively. Based on basic principle of theoretical mechanics, principal vector of force system acting on center of mass is

$$R = \sum_{i=1}^{4} F_{i,R} + F_d \tag{3.89}$$

It shall be noted that F_d acts on the origin, has impact only on principal vector and generates no couple. Therefore, principal moment of force system acting on center of mass is

$$M_o = \sum_{i=1}^{4} r_i \times F_{i,R} \tag{3.90}$$

Combine Eqs. (3.89) and (3.90), and then

$$Q = \sum_{i=1}^{4} \begin{bmatrix} F_{i,R} \\ g_i^T F_{i,R} \end{bmatrix} + \begin{bmatrix} F_d \\ 0 \end{bmatrix} \tag{3.91}$$

wherein, $g_i = [-b_i, l_i]^T$ relies on structure of vehicles and $Q = [F_x, F_y, M]^T$. Acceleration a_x, a_y acting on center of mass of vehicle (origin of vehicle coordinate system) and acceleration of yaw angle $\dot{\gamma}$ can be obtained by momentum theorem and theorem of moment of momentum [30], and then dynamic equation of full vehicle can be established as

$$U = CQ \tag{3.92}$$

wherein, $U = [a_x, a_y, \dot{\gamma}]^T, C = \text{diag}\{m^{-1}, m^{-1}, I_z^{-1}\}$. INS in measurement equation shown in Eq. (3.49) can be acquired by Eq. (3.92).

(3) Analysis on vehicle kinematic model
Motion situations of vehicles can be described by longitudinal velocity v_x, lateral velocity v_y and yaw rate γ generally, which are several state vectors of most concern in vehicle dynamic control field. Relation between lateral acceleration and state vector v_x, v_y and γ can be obtained according to kinematic relation of vehicles [31], as shown in Eq. (3.93).

$$\begin{cases} a_x = \dot{v}_x - v_y\gamma \\ a_y = \dot{v}_y + v_x\gamma \end{cases} \tag{3.93}$$

Define key motion state of vehicles as vector $V = [v_x, v_y, \gamma]^T$. In consideration of kinematic relation of vehicles, solve Eqs. (3.92) and (3.93) simultaneously and then the following Equation can be obtained:

$$\dot{V} = U + G(V)V \tag{3.94}$$

wherein,

$$G(V) = \begin{bmatrix} 0 & -\gamma & 0 \\ \gamma & 0 & 0 \\ 0 & 0 & 0 \end{bmatrix} \tag{3.95}$$

Solve Eqs. (3.84)–(3.95) simultaneously and then a complete set of dynamic equation of full vehicle can be established. Wherein, motion state of vehicles in state recurrence equation (3.48) can be established by Eq. (3.94).

(4) Analysis on tire kinematic model
As shown in Fig. 3.10, velocity of driving wheel center is $v_{i,R} = [v_{i,X}, v_{i,Y}]^T$ in coordinate system $x_R y_R$ and transformational matrix from state vector V to velocity of driving wheel center $v_{i,V}$ is $P_{i,V}$.

$$v_{i,V} = P_{i,V}V \tag{3.96}$$

$$P_{i,V} = [I_2 \quad g_i] \tag{3.97}$$

Velocity of driving wheel center is $v_i = [v_{i,x}, v_{i,y}]^T$ in wheel coordinate system $x_W y_W$. From Eq. (3.85), it can been seen that transformational matrix from coordinate system $x_W y_W$ to $x_R y_R$ is $\boldsymbol{R}(\delta_i)$. Given that rotation transformational matrixes are all orthogonal matrixes, transformational matrix from $x_R y_R$ to $x_W y_W$ is $\boldsymbol{R}^T(\delta_i)$. Thus,

$$v_i = \boldsymbol{R}^T(\delta_i)v_{i,R} \tag{3.98}$$

According to Eq. (3.98), velocity of wheel center in rolling direction of tire is

$$v_{i,x} = \boldsymbol{L}_i^T V \tag{3.99}$$

wherein,

$$\boldsymbol{L}_i = [\cos\delta_i, \ \sin\delta_i, \ -b_i\cos\delta_i + l_i\sin\delta_i]^T \tag{3.100}$$

3.4.3 Analysis on Magic Formula Tire Model

In consideration of trackslip situation of vehicle wheels, computational Eq. (3.101) [28] for longitudinal trackslip rate κ_i can be obtained by Magic Formula unified trackslip Equation.

$$\kappa_i = \frac{R_i\omega_i - v_{i,x}}{v_{i,x}} \tag{3.101}$$

wherein, R_i refers to the rolling radius of tire, ω_i refers to the rotate velocity of vehicle wheels and $v_{i,x}$ refers to the velocity of wheel center in rolling direction of tires. Computational Equation of slip angle of tires is

$$\alpha_i = \arctan\frac{v_{i,X}}{v_{i,Y}} - \delta_i \tag{3.102}$$

Since the acting force between tires and road surface is impacted by various factors, non-linear property of tires is obvious. Such factors include vehicle wheels itself (such as tire pressure, tire thread and wearing degree) and external factors (such as vertical load of tires, slip angle of tires and slip rate of tires). Various non-linear tire models have been set home and abroad, such as Brush model [28], Dugoff model [32] and Uni-Tire model [33, 34].

Magic Formula Model was proposed by Volvo Company and Delft University of Netherlands, in which Professor Pacejka played the most significant role [28]. Magic Formula Model is a semi-empirical model based on measured data, which is applicable to various tire structures and different working conditions and is the most accurate one in various widely-used tire models. Application of Magic Formula

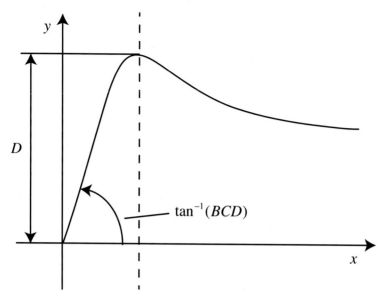

Fig. 3.11 Magic formula model: curve of tire property

Model in state estimation of vehicles can provide more accurate information. Hence, Magic Formula Model was adopted in this dissertation for discussion of State Estimation. Figure 3.11 is the characteristic curve of tire drawn by making use of Magic Formula Model.

For improving computing speed and simplifying computing flow, simplification shall be performed for Magic Formula Model. The basic assumptions include that road surface adhesion coefficient of road surface and rolling radius of tire have been known [10], and that aligning torque and camber angle of vehicle wheels can be neglected. On the basis of the above assumptions, Magic Formula Model is simplified as follows [28].

$$y = D \sin[C \arctan\{Bx - E(Bx - \arctan(Bx))\}] \tag{3.103}$$

See Table 3.1 for represented meanings of various parameter variables in Eq. (3.103).

Computing flow of Magic Formula Model is as follows.

(1) Under the condition of pure sideslip and side trackslip, longitudinal force and lateral force decouple and then calculate nominal longitudinal force F_{x0} and nominal lateral force F_{y0};

(2) During co-occurrence of side slip and trackslip, longitudinal force and lateral force impacts each other interactively and then calculate the impact factor $G_{x\alpha}$ of side slip on longitudinal force and impact factor $G_{y\kappa}$ of trackslip on lateral force;

Table 3.1 Parameter and variable in magic formula model

Parameter	Definition
x	Longitudinal slip ratio κ and tire slip angle α
y	Longitudinal force F_x and traversal force F_y
B	Stiffness coefficient
C	Shape coefficient
D	Point of max value
E	Coefficient of graduation

(3) During co-occurrence of side slip and trackslip, calculate longitudinal force F_x and lateral force F_y of tires. Calculating parameter of nominal longitudinal force and lateral force are as shown in Table 3.2.

In Table 3.2, parameter p_j and q_j depend on road surface adhesion coefficient μ and tire pressure P_{tire}, which are considered to be known. However, new-type observation method of road surface adhesion coefficient was not discussed in this dissertation, but estimated value of road surface adhesion coefficient was obtained directly from previous research results [8]. Hence, p_j and q_j can be obtained from the table. Meanwhile, as one of the important variables, vertical force F_z has been obtained in Sect. 3.3.

Calculating parameter of nominal longitudinal force and lateral force are as shown in Table 3.2.

In Table 3.2, parameter p_j and q_j depend on road surface adhesion coefficient μ and tire pressure P_{tire}, which are considered to be known. However, new-type observation method of road surface adhesion coefficient was not discussed in this dissertation, but estimated value of road surface adhesion coefficient was obtained directly from previous research results [10]. Hence, p_j and q_j can be obtained from the table. Meanwhile, as one of the important variables, vertical force F_z has been obtained in Sect. 3.3 ("Vertical Force Estimation of Tire Based on Multi-Information Fusion").

Table 3.2 Calculation of nominal longitudinal force and lateral force

Parameter	Nominal longitudinal force	Nominal lateral force
x	κ	α
y	F_{x0}	F_{y0}
C	p_0	q_0
D	$p_1 F_z^2 + p_2 F_z$	$q_1 F_z^2 + q_2 F_z$
BCD	$(p_3 F_z^2 + p_4 F_z)e^{p_5 F_z}$	$q_3 \sin(q_4 \arctan(q_5 F_z))$
E	$p_6 F_z^2 + p_7 F_z + p_8$	$q_6 F_z + q_7$

Table 3.3 Calculation of longitudinal and lateral impact factor

Parameter	Longitudinal impact factor	Lateral impact factor
x	κ	α
y	$G_{x\alpha}$	$G_{y\kappa}$
C	$C_{x\alpha}$	$C_{y\kappa}$
B	$B_{x\alpha}$	$B_{y\kappa}$
S_H	$S_{H,x\alpha}$	$S_{H,y\kappa}$

Impact factors $G_{x\alpha}$ and $G_{y\kappa}$ can be calculated by Eq. (3.104), which is simplified from reference [28].

$$y = \frac{\cos(C \arctan(B(x + S_H)))}{\cos(C \arctan(B S_H))} \tag{3.104}$$

In Eq. (3.104), represented meanings of each parameter are as shown in Table 3.3. In Table 3.3, parameter $B_i C_i S_{H,i}$, $i \in \{x\alpha, y\kappa\}$ can be acquired by substituting actual data into Eq. (3.104) for re-matching. Longitudinal force F_x of tire can be calculated and obtained by Eq. (3.105).

$$F_x = G_{x\alpha} F_{x0} \tag{3.105}$$

Under the condition of quasi-static state, lateral force $\overline{F}_y$ of tire can be calculated and obtained by Eq. (3.106).

$$\overline{F}_y = G_{y\kappa} F_{y0} \tag{3.106}$$

As shown in Fig. 3.11 and characteristic curve of tire drawn by Magic Formula Model, if longitudinal force of tire does not exceed 80 % of the maximum adhesion force that road surface can be offered, longitudinal trackslip rate κ and longitudinal force F_x are in basic direct proportion [35]. In such a case, take longitudinal trackslip rate as variable and solve partial derivations of longitudinal force in origin and initial longitudinal slip stiffness $C_{\kappa 0}$ as

$$C_{\kappa 0} = G_{x\alpha} B C D \tag{3.107}$$

Under normal driving situations, estimated value $\hat{\kappa}$ of longitudinal trackslip rate can be expressed by Eq. (3.108).

$$\hat{\kappa} \doteq \frac{F_x}{C_{\kappa 0}}, \quad F_x \leq 80 \% G_{x\alpha} D \tag{3.108}$$

Solve Eqs. (3.101) and (3.108) simultaneously and then rotate velocity ω_i of driving wheels can be obtained.

$$\dot{\omega}_i \doteq \frac{(1+\hat{\kappa})V_{i,x}}{R_i} \tag{3.109}$$

If the trackslip rate is relatively smaller, Eq. (3.109) can approximate wheel velocity well. In comprehensive consideration of Eqs. (3.99) and (3.109), wheel velocity in measurement equation shown in Eq. (3.49) can be obtained.

3.4.4 Analysis on Dynamic Tire Model

For the calculation of lateral force, simplified Magic Formula Model adopted in Sect. 3.4.3 is applicable to quasi-static state working conditions only. When velocity of vehicles changes, dynamic tire model shall be introduced [35]. A commonly-used dynamic tire model is as shown in Eq. (3.110).

$$\dot{F}_y = \frac{K_0 V_x}{C_{\alpha 0}}(\overline{F}_y - F_y) \tag{3.110}$$

$$C_{\alpha 0} = \left.\frac{\partial F_y}{\partial \alpha}\right|_{\alpha=0} \tag{3.111}$$

$$K_0 = \left.\frac{\partial F_y}{\partial y}\right|_{y=0} \tag{3.112}$$

wherein, $\overline{F}_y$ is the estimated result of lateral force under quasi-static working conditions and obtained according to Eq. (3.106). V_x refers to the longitudinal velocity of wheel center; $C_{\alpha 0}$ refers to the initial cornering stiffness of tire; K_0 refers to the initial lateral displacement stiffness of tire; y refers to the sideslip displacement of wheel center against road surface. Equation (3.110) is applicable to description of lateral force of tire under changing vehicle velocity [36, 37] and can be applied to establishing the part of tire lateral force in state recurrence equation (3.48).

3.4.5 Calibration of Inertial Sensor

Sampling frequency of INS is relatively higher in generation, which can be as high as 100 Hz. Main defect of INS is its vulnerability to be disturbed by external environment, as a result of which static error often arises. Performing numerical integration by direct use of measured value of INS and then estimation of motion state of vehicles tends to lead to relatively larger deviation. Hence, direct use of INS is not allowed to perform long-time numerical integration [38].

With the development of navigation technology of vehicles, GPS is integrated in more and more full vehicle system. Assisting vehicle State Estimation by GPS information has gradually become one of hot points for study [1, 2, 38–49]. The advantages of GPS lie in its accurate acquisition of instantaneous velocity without the need of numerical integration, since Doppler effect is applied in GPS velocity estimation [49]. Major defect of GPS is the low sampling frequency. Sampling frequency of in-vehicle GPS system is 1 Hz in general, which cannot be directly used for vehicle dynamic control.

In combination of high-precision and velocity of GPS and high sampling frequency of INS and overcoming defects of the two, performing observation of state vehicle parameters by integration of GPS and INS has become a hot point of study. Discussion is made in reference [44–46] about integration of GPS and INS and majority of GPS sensors adopted in these references is 10 Hz or so. The functional mechanism is that when GPS signal exists at sampling moment, velocity information of GPS is used at this moment and INS static deviation is estimated; when no GPS signal exists at sampling moment, INS signal integration after deviation removal is adopted. Given relatively short integration time in this method (10 steps, 0.1s), integration deviation has little impact on observation of state parameter. However, if common 1 Hz GPS is adopted to integrate INS information, larger outcome integration deviation may arise due to too much integration steps (100 steps, 1s). Therefore, only relatively accurate velocity information obtained by 1 Hz GPS was adopted in this dissertation to calibrate inertial sensor so as to facilitate further State Estimation. Figure 3.12 shows the vehicle coordinate system, INS coordinate system and geodetic coordinate system, which can be used to describe relation among various coordinate systems in this section. Yaw rate sensor was firstly calibrated in this dissertation and on such a basis longitudinal/lateral acceleration sensor was further calibrated.

(1) Calibration of yaw rate sensor

Define included angle between vertical axis of vehicle and due east as the direction angle ψ_V of vehicles to represent the current attitude of vehicles. Define included angle between center of mass velocity of vehicle and due east as the course angle ψ_{GPS} of vehicle to present actual heading direction of vehicles. Relation among course angle, direction angle and side slip angle is as shown in Eq. (3.113).

$$\psi_{GPS} = \psi_V + \beta \tag{3.113}$$

Measurement value of yaw rate is γ_m, which can be considered to be the result of joint action of actual yaw rate γ and static deviation γ_b of yaw rate, as shown in Eq. (3.114).

$$\gamma_m = \gamma + \gamma_b \tag{3.114}$$

Static error of yaw rate can be deemed as constant in a short period, as a result of which change rate of static error of yaw rate can be regarded as 0, as shown in Eq. (3.115).

$$\dot{\gamma}_b = 0 \tag{3.115}$$

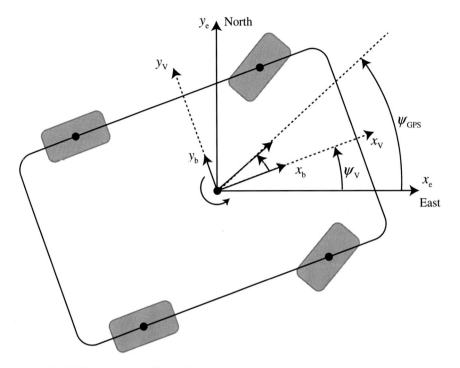

Fig. 3.12 Calibration model of inertial sensor

When straight driving velocity or yaw rate of vehicles is relatively small, it can be considered that side slip angle β at this moment is very small and direction angle is approximate to course angle, as shown Eq. (3.117).

$$\psi_{GPS} \doteq \psi_V + r_{\psi_{GPS}} \tag{3.116}$$

When straight driving velocity or yaw rate of vehicles is relatively small, it can be considered that side slip angle β at this moment is very small and direction angle is approximate to course angle, as shown Eq. (3.117).

$$\psi_{GPS} \doteq \psi_V \tag{3.117}$$

When driving direction of vehicles steers or yaw rate is relatively larger, condition of Eq. (3.117) is no longer met. At this moment, there is no need to calibrate yaw rate sensor. According to Eqs. (3.114)–(3.117), state observer is designed as

$$\begin{cases} \dot{x} = Ax + Bu + q \\ y = Cx + r \end{cases} \tag{3.118}$$

$$x = \begin{bmatrix} \psi_V \\ \gamma_b \end{bmatrix}, A = \begin{bmatrix} 0 & -1 \\ 0 & 0 \end{bmatrix}, B = \begin{bmatrix} 1 \\ 0 \end{bmatrix}, u = \gamma_m, y = \psi_{GPS}$$

$$C = \begin{cases} [0\ 1], & \text{When straightly drive or yaw rate is lower} \\ [0\ 0], & \text{When steers or yaw rate is higher} \end{cases}$$

wherein, state vector to be observed is x, q refers to the process noise; r refers to the measurement noise. Covariance matrix of process noise is Q and that of measurement noise is R. Deviation γ_b of yaw rate can be satisfactorily estimated by making use of Kalman Filtering shown in Sect. 3.3.1 so as to obtain relatively accurate yaw rate information from sensor signals.

(2) Calibration of longitudinal/lateral acceleration sensor

The advantage of longitudinal/lateral acceleration sensor lies in relatively higher sampling frequency. However, given that longitudinal/lateral acceleration sensor tends to be impacted by temperature, vibration and installation position, there exists static deviation generally. Thus, it is unable to apply direct integration of longitudinal/lateral acceleration sensor information to obtain velocity information of vehicles and calibration for sensor is necessary. Before calibrating of longitudinal/lateral acceleration sensor, side slip angle needs to be determined initially. Given that GPS is adopted in this method to perform initial estimation of side slip angle, side slip angle acquired by this method is defined as GPS. Given that measuration frequency of β_{GPS} is 1 Hz and cannot be directly used for vehicle dynamic control, β_{GPS} serves as one intermediate variable for calibration of acceleration sensor; the computing process for β_{GPS} is as follows. When yaw rate of vehicles is very small, side slip angle of vehicles can be regarded as very small, i.e.

$$\beta_{GPS} = 0 \tag{3.119}$$

Supposing starting from t_0, yaw rate of vehicles exceeds the set value and side slip angle of vehicles is no longer considered to be 0, then side slip angle of vehicle at t moment can be estimated as

$$\beta_{GPS}(t) = \psi_{GPS}(t) - \psi_{GPS}(t_0) - \int_{t_0}^{t} \gamma(t)dt \tag{3.120}$$

According to Eqs. (3.119) and (3.120), initial estimated value β_{GPS} of side slip angle can be obtained. Relation between acceleration sensor and vehicle motion state can be acquired based on vehicle dynamics principle [31], calculation method of which is as follows:

$$\begin{cases} a_{x,m} = \dot{v}_{x,GPS} - v_{y,GPS}\gamma + a_{x,b} \\ a_{y,m} = \dot{v}_{y,GPS} + v_{x,GPS}\gamma + a_{y,b} \end{cases} \tag{3.121}$$

wherein, $a_{x,m}$ refers to the measurement value of longitudinal acceleration sensor; $a_{y,m}$ refers to the measuration value of lateral acceleration sensor; $a_{x,b}$ refers to the static deviation of longitudinal acceleration sensor; $a_{y,b}$ refers to the static deviation of lateral acceleration sensor. Static deviation of longitudinal/lateral acceleration sensor can be regarded as constant within a short period, thus

$$\begin{cases} \dot{a}_{x,b} = 0 \\ \dot{a}_{y,b} = 0 \end{cases} \tag{3.122}$$

Longitudinal/lateral velocity of vehicles can be obtained as follows by velocity v_{GPS} acquired with initially-estimated side slip angle β_{GPS} and GPS sensor.

$$\begin{cases} v_{x,GPS} = v_{GPS} \cos \beta_{GPS} \\ v_{y,GPS} = v_{GPS} \sin \beta_{GPS} \end{cases} \tag{3.123}$$

State observer can be designed as follows according to Eqs. (3.121) to (3.123).

$$\begin{cases} \dot{x} = Ax + Bu + q \\ y = Cx + r \end{cases} \tag{3.124}$$

$$x = \begin{bmatrix} v_{x,GPS} \\ a_{x,b} \\ v_{y,GPS} \\ a_{y,b} \end{bmatrix}, A = \begin{bmatrix} 0 & -1 & \gamma & 0 \\ 0 & 0 & 0 & 0 \\ -\gamma & 0 & 0 & -1 \\ 0 & 0 & 0 & 0 \end{bmatrix}, B = \begin{bmatrix} 1 & 0 \\ 0 & 0 \\ 0 & 1 \\ 0 & 0 \end{bmatrix}$$

$$u = \begin{bmatrix} a_{x,m} \\ a_{y,m} \end{bmatrix}, y = \begin{bmatrix} v_{x,GPS} \\ v_{y,GPS} \end{bmatrix}, C = \begin{bmatrix} 1 & 0 & 0 & 0 \\ 0 & 0 & 1 & 0 \end{bmatrix}$$

wherein, q refers to the process noise; r refers to the measurement noise; covariance matrix of process noise is Q and that of measurement noise is R. Deviation $a_{x,b}$ and $a_{y,b}$ of longitudinal/lateral acceleration can be estimated satisfactorily by making use of Kalman Filtering shown in Sect. 3.3.1 and relatively accurate information concerning longitudinal/lateral acceleration can be obtained from sensor signals. Ultimately, process yaw rate, measured value of longitudinal/lateral acceleration sensor into calibrated corrected value, as shown from Eqs. (3.125) to (3.127).

$$\gamma = \gamma_m - \gamma_b \tag{3.125}$$

$$a_x = a_{x,m} - a_{x,b} \tag{3.126}$$

$$a_y = a_{y,m} - a_{y,b} \tag{3.127}$$

Corrected value of inertial sensor after calibration can be used for INS of measurement vector y in unscented particle filter.

3.4.6 Self-adaptive Adjustment of Measurement Noise

As shown in Eq. (3.47), measurement vector y is composed of corrected value of inertial sensor and rotate velocity of each wheel. Supposing that the information obtained from each measuration is independent of one another, then covariance matrix R of measurement noise can be simplified as diagonal matrix, as shown Eq. (3.128).

$$R = \mathrm{diag}\{r_{a_x}, r_{a_y}, r_\gamma, r_{\omega 1}, r_{\omega 2}, r_{\omega 3}, r_{\omega 4}\} \tag{3.128}$$

Although covariance matrix of noise cannot be given accurately, information of vehicles and covariance matrix of noise is close related, according to which the following self-adaptive adjustment principle of measurement noise is given.

(1) When signal a_x, a_y, γ of inertial sensor is relatively small, signal-to-noise ratio of inertial sensor is relatively small and reliability is lowered, measurement noise $r_{a_x}, r_{a_y}, r_\gamma$ of inertial sensor shall be increased;

(2) When longitudinal driving force of wheels is relatively large and within the non-linear interval shown in Fig. 3.11, trackslip rate of wheels is relatively large and reliability of wheel velocity information is lowered, noise r_{ω_i} of wheel velocity information shall be increased and noise r_{a_x} of acceleration information shall be reduced at the same time;

(3) When steering angle δ of steering wheel is relatively small, yaw rate γ shall be relatively small and corresponding noise r_γ to yaw rate shall be increased;

(4) When steering angle δ of steering wheel is relatively small, lateral acceleration a_y shall be relatively small and corresponding noise r_{a_y} to lateral acceleration shall be increased.

In consideration of the above principle, adoption of RISF method to design covariance matrix will expect sound effect [50]. Here, the information of various vehicles is adopted to design measurement noise, as shown from Eqs. (3.129) to (3.132).

$$r_{a_x} = c_{a_{x1}} \exp(-d_{a_{x1}}|a_x|) + c_{a_{x2}} \exp(-d_{a_{x2}}|\sum_{i=1}^{4} F_{i,x}|) + c_{a_{x3}} \exp(-d_{a_{x3}} \sum_{i=1}^{4} |\hat{\kappa}_i|)$$
$$\tag{3.129}$$

$$r_{a_y} = c_{a_{y1}} \exp(-d_{a_{y1}}|a_y|) + c_{a_{y2}} \exp(-d_{a_{y2}}|\delta|) \tag{3.130}$$

$$r_\gamma = c_{\gamma_1} \exp(-d_{\gamma_1}|\gamma|) + c_{\gamma_2} \exp(-d_{\gamma_2}|\delta|) \tag{3.131}$$

$$r_{\omega_i} = c_{\omega_{i1}} \exp(d_{\omega_{i1}}|\hat{\kappa}_i|) + c_{\omega_{i2}} \exp(d_{\omega_{i2}}|a_x|) + c_{\omega_{i3}} \exp(d_{\omega_{i3}}|a_y|) \tag{3.132}$$

wherein, c_j, d_j are arithmetic number, $j \in \{a_{x1}, a_{x2}, a_{x3}, a_{y1}, a_{y2}, \gamma_1, \gamma_2, \omega_{i1}, \omega_{i2}, \omega_{i3}\}$, $i \in \{1, 2, 3, 4\}$. The designed covariance matrix R of measurement noise will be applied to self-adaptive adjustment of measurement noise of unscented particle filter as shown in Eq. (3.53), which can comprehensively improve observation precision under various conditions and promotes anti-interference.

3.5 Brief Summary

This chapter studied state estimation in dynamic control of distributed electric vehicles. Compared with traditional vehicles, accurate information concerning in-wheel driving force and wheel velocity were added for distributed electric vehicles. On such a basis and in combination of GPS, INS and driver's operation information, a complete set of state estimation system for distributed electric vehicles was formed in this chapter, with accurate information of in-wheel driving force being adopted for quality estimation and gradient estimation. However, in estimation of motion state and lateral force of vehicles, multiple kinds of information was fully used, such as information of in-wheel driving force, wheel velocity information, GPS and INS.

State observation methods for mass of full vehicle was discussed firstly in this chapter, with quality estimation method based on high-frequency information extraction being proposed, which decouples quality estimation and road surface gradient, effectively removing the relatively great impact of road surface gradient on quality estimation. Ultimately, sound quality estimation result was acquired by making use of RLS. After acquisition of relatively accurate quality estimation result, kinematic and dynamic methods were adopted to perform multi-method combined observation on road surface gradient. In terms of dynamics, time-variant characteristic of gradient was taken into consideration, and RLS with forgetting factor was adopted to make gradient estimation, while in kinematics, strong correlation between gradient and static deviation of sensor was adopted to obtain the current gradient directly. By coordination and complementation of dynamics and kinematics, heavy reliance on vehicle model precision by gradient estimation and limitation of great impact on static error of acceleration sensor were solved, as a result of which rapid and efficient estimation could be made under various working conditions. After quality information and gradient information were obtained, observation method of tire vertical force based on multi-information fusion was put forward. This observation method of vertical force integrated information of multiple sources, such as the information of longitudinal/lateral acceleration, the estimated quality and gradient information as well as the information of roll angle velocity sensor. On the basis of estimation of dynamic axle load transfer and roll angle, vertical force of each wheel was accurately observed. Observation precision improvement of vertical force will further improve observation precision of state parameter of other vehicles, ending observation result distortion of other states caused by observation error of vertical force.

On the basis of acquiring mass of full vehicle, road surface gradient and vertical force of tire, combined observation method of state parameter of vehicles based on unscented particle filter was put forward, while longitudinal velocity, side slip angle and yaw rate of vehicles and lateral force of wheels were observed. Meanwhile, for tackling static deviation of INS, in-vehicle GPS was utilized to calibrate INS and INS static deviation was removed in real time, laying sound foundation for follow-up use of INS information. In addition, in observation study of lateral force, Magic Formula Model was proposed with dynamic property being taken into consideration and efficient estimation on lateral force of tire was conducted in motion process. Through self-adaptive adjustment of measurement noise, effective fusion of multi-sensor information was realized, improving the applicable range of the state observation algorithm. Owing to comprehensive utilization of various types of information offered by distributed electric vehicles, state estimation methods proposed in this dissertation will be able to satisfy the observation precision demand under various working conditions.

References

1. Bae HS, Gerdes JC (2000) Parameter estimation and command modification for longitudinal control of heavy vehicles. In: Proceedings of the international symposium on advanced vehicle control, Ann Arbor, Michigan, USA
2. Sahlholm P, Johansson KH (2010) Road grade estimation for look-ahead vehicle control using multiple measurement runs. Control Eng Pract 18(11):1328–1341
3. Lingman P, Schmidtbauer B (2002) Road slope and vehicle mass estimation using Kalman filtering. Veh Syst Dyn 37(Supplement):12–23
4. Madsen CK, Zhao JH (1999) Optical filter design and analysis: a signal processing approach. Wiley, New York
5. Gibbs BP (2011) Advanced Kalman filtering, least-squares and modeling. Wiley, Hoboken
6. Åström KJ, Wittenmark B (2008) Adaptive control, 2nd edn. Dover Publications, Mineola
7. Parkum JE, Poulsen NK, Holst J (1992) Recursive forgetting algorithms. Int J Control 55(1):109–128
8. Saelid S, Foss B (1983) Adaptive controllers with a vector variable forgetting factor. Proceedings of the IEEE conference on decision and control. San Antonio, Texas, USA, Dec 1983, pp 1488–1494
9. Piyabongkarn D, Rajamani R, Grogg JA et al (2009) Development and experimental evaluation of a slip angle estimator for vehicle stability control. IEEE Trans Control Syst Technol 17(1):78–88
10. Liu L, Luo Y, Li K (2009) Observation of road surface adhesion coefficient based on normalized tire model. J Tsinghua Univ (Nat Sci) 49(5):116–120
11. Kalman RE (1960) A new approach to linear filtering and prediction problems. Trans ASME J Basic Eng 82(Series D):35–45
12. Welch G, Bishop G (2006) An introduction to the Kalman filter. Technical report, Department of Computer Science, University of North Carolina at Chapel Hill. http://www.cs.unc.edu/~welch/kalman/
13. van Zanten AT (2000) Bosch ESP systems: 5 years of experience. SAE technical paper: 2000-01-1633
14. Stphant J, Charara A, Meiz D (2004) Virtual sensor: application to vehicle sideslip angle and transversal forces. IEEE Trans Ind Electron 51(2):278–289
15. Kalman RE, Bucy RS (1961) New results in linear filtering and prediction theory. Trans ASME J Basic Eng 83(Series D):95–108

16. Yu Z, Gao X (2009) Review of vehicle state estimation problem under driving situation. J Mech Eng 45(5):20–33
17. Haykin S (2001) Kalman filtering and neural networks. Wiley, New York
18. Julier SJ, Uhlmann JK (1997) A new extension of the Kalman filter to nonlinear systems. In: Proceedings of the international symposium on aerospace/defense sensing, simulation and controls. Orlando, Florida, USA, Apr 1997, pp 182–193
19. Julier SJ, Uhlmann JK (2004) Unscented filtering and nonlinear estimation. Proc IEEE 92(3):401–422
20. Wan EA, van der Merwe R (2000) The unscented Kalman filter for nonlinear estimation. In: Proceedings of the IEEE adaptive systems for signal processing, communications, and control symposium. Lake Louise, Alberta, Canada, Oct 2000, pp 153–158
21. Chen Z (2003) Bayesian filtering: from Kalman filters to particle filters, and beyond. Technical report, McMaster University, Hamilton
22. Carpenter J, Clifford P, Fearnhead P (1999) Improved particle filter for nonlinear problems. IEE Proc Radar Sonar Navig 146(1):2–7
23. Arulampalam MS, Maskell S, Gordon N et al (2002) A tutorial on particle filters for online nonlinear/non-Gaussian Bayesian tracking. IEEE Trans Signal Process 50(2):174–188
24. van der Merwe R, Doucet A, de Freitas N et al (2000) The unscented particle filter. Technical report, Engineering Department, Cambridge University, Cambridge
25. Crisan D, Doucet A (2002) A survey of convergence results on particle filtering methods for practitioners. IEEE Trans Signal Process 50(3):736–746
26. Julier SJ, Uhlmann JK, Furrant-Whyten HF (1995) A new approach for filtering nonlinear systems. In: Proceedings of the American control conference. Seattle, Washington, USA, June 1995, pp 1628–1632
27. Deng X, Xie J, Guo W (2006) Adaptive particle filtering based on state estimation. J South China Univ Technol (Nat Sci) 34(1):57–61
28. Pacejka HB (2012) Tire and vehicle dynamics, 3rd edn. Elsevier, Oxford
29. Kiencke U, Nielsen L (2010) Automotive control systems: for engine, driveline, and vehicle, 2nd edn. Springer, New York
30. Li J, Zhang X (2010) Theoretical mechanics, No. 2 Version. Tsinghua University Press, Beijing
31. Yu Z (2009) Automobile theory, No. 5 Version. Tsinghua University Press, Beijing
32. Dugoff H, Fancher PS, Segel L (1970) The influence of lateral load transfer on directional response. SAE technical paper 700377
33. Guo K, Ren L (1999) A unified semi-empirical tire model with higher accuracy and less parameters. SAE technical paper: 1999–01–0785
34. Guo K (2011) Principles of vehicle control dynamics. No. 3 Version. Jiangsu Science and Technology Press, Nanjing
35. Rajamani R (2012) Vehicle dynamics and control, 2nd edn. Springer, New York
36. Loeb JS, Guenther DA, Chen FH (1990) Lateral stiffness, cornering stiffness and relaxation length of the pneumatic tire. SAE technical paper 900129
37. Heydinger GJ, Garrott WR, Chrstos JP (1991) The importance of tire lag on simulated transient vehicle response. SAE technical paper 910235
38. Leung KT, Whidborne JF, Purdy D et al (2011) A review of ground vehicle dynamic state estimations utilising GPS/INS. Veh Syst Dyn 49(1):29–58
39. Bae HS, Ryu J, Gerdes JC (2001) Road grade and vehicle parameter estimation for longitudinal control using GPS. In: Proceedings of the IEEE conference on intelligent transportation systems. Oakland, California, USA, Aug 2001
40. Johansson K (2005) Road slope estimation with standard truck sensors. KTH Royal Institute of Technology, Sweden, Apr 2005
41. Jansson H, Kozica E, Sahlholm P et al (2006) Improved road grade estimation using sensor fusion. In: Proceedings of the 12th Reglermöte, Stockholm, Sweden, May 2006
42. Parviainen J, Hautamäki J, Collin J et al (2009) Barometer-aided road grade estimation. In: Proceedings of the world congress of the international association of institutes of navigation. Stockholm, Sweden, Oct 2009

43. Dai Y, Luo Y, Chu W et al (2012) Vehicle state estimation based on the integration of low-cost GPS and INS. In: Proceedings of the international conference on advanced vehicle technologies and integration. Changchun, China, July 2012, pp 677–681
44. Bevly DM (2004) Global Positioning System (GPS): a low-cost velocity sensor for correcting inertial sensor errors on ground vehicles. J Dyn Syst Meas Control 126(2):255–264
45. Bevly DM, Gerdes JC, Wilson C (2002) Use of GPS based velocity measurements for measurement of sideslip and wheel slip. Veh Syst Dyn 38(2):127–147
46. Ryu J (2004) State and parameter estimation for vehicle dynamics control using GPS. Stanford University, USA, Dec 2004
47. Zhang T, Yang D, Li T et al (2010) Vehicle state estimation system aided by inertial sensors in GPS navigation. In: Proceedings of the international conference on electrical and control engineering. Wuhan, China, June 2010
48. Zhang T (2010) Behavior matching of vehicle driving roads. Tsinghua University, Beijing
49. Grewal M, Weill L, Andrewsa A (2007) Global positioning systems, inertial navigation, and integration. Wiley, Hoboken
50. Lee H (2006) Reliability indexed sensor fusion and its application to vehicle velocity estimation. J Dyn Syst Meas Control 128(2):236–243

Chapter 4
Coordinated Control of Distributed Electric Vehicles

Abstract The state observation method for distributed electric drive vehicle described in Chap. 3 can provide the support for the dynamic control of vehicle. The distributed electric drive vehicle is equipped with multiple driving wheels without mechanical connection, and the property of distributed drive can ensure the control system to perform independent control allocation on the driving force of each wheel. The reasonable utilization of properties of distributed electric drive vehicle is significant to the comprehensive optimizing of dynamic control property of full vehicle on the premises of meeting drivers' requirements and ensuring safety. According to the properties of the distributed electric drive vehicle, this dissertation designed a complete set of coordinated control system for distributed electric drive vehicle, including three technologies, i.e. determination of vehicle dynamic demand target, driving force control allocation, and motor property compensation control. For the determination of vehicle dynamic demand target, tracing of desired side slip angle and desired yaw rate is taken as the lateral stability control target. The direct yaw moment method is used for controlling the lateral stability of vehicle, so as to enhance the safety of vehicles. The expected yaw rate and side slip angle are made by the non-linear vehicle model, and the expectation value is taken as the lateral stability control target, and the scheme which comprehensively considers the $\beta - \dot{\beta}$ phase diagram and road surface adhesion coefficient is put forward when the weights of the yaw rate and side slip angle are adjusted. For driving force control allocation, many limitation conditions required by control allocation of distributed electric drive vehicle were taken into consideration, including motor driving capacity, motor failure condition, road surface adhesion condition, etc. When the allocation is optimized, the motor utilization rate is defined for describing the utilization condition of distributed electric drive vehicle, and the optimum objective function is defined to comprehensively reduce the utilization rate of all motors and protect the failed motor. When optimizing, the designed objective function is solved by the functional collection method, so as to further perform control allocation on the driving force of each wheel. For motor property compensation control, the dynamic response error and steady-state response error between different motors are taken into consideration, the self-adaptive control method is adopted for correcting the dynamic response and steady-state response properties of all motors, then the responses of all motors are approximately consistent, and the command on control allocation layers is accurately

© Springer-Verlag Berlin Heidelberg 2016
W. Chu, *State Estimation and Coordinated Control for Distributed Electric Vehicles*, Springer Theses, DOI 10.1007/978-3-662-48708-2_4

completed in real time. The three technologies respectively determine the demand target of full vehicle, the driving force of each wheel is subject to control allocation, the moment command on driving wheels is made to execute the moment allocation value of each wheel, and then a complete set of coordinated control system is formed, so as to meet the demand target of full vehicle.

4.1 Determination of Vehicle Dynamic Demand Target

This section will define the demand of full vehicle according to the drivers' operation and current running state of vehicle, including desired longitudinal driving force and desired direct yaw moment. The algorithm process of determination of vehicle dynamic demand target is as shown in Fig. 4.1.

4.1.1 Determination of Desired Longitudinal Driving Force

As the driver self has the obvious feedback control effect on longitudinal driving, and the coupling relationship with lateral safety is not obvious, the designed desired longitudinal driving force is open-loop, and the closed-loop control is completed by the driver, as shown in Eq. (4.1).

$$F_{x,\mathrm{d}} = K_{a\mathrm{p}} \alpha_{\mathrm{P}} \tag{4.1}$$

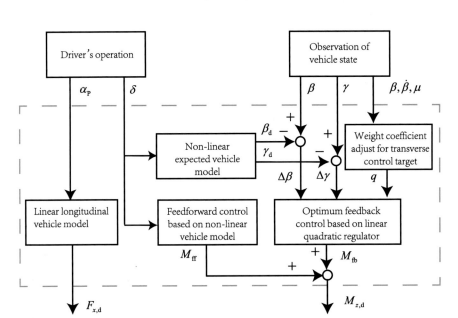

Fig. 4.1 Determination of vehicle dynamic demand target

wherein, α_P refers to the opening degree of acceleration pedal, K_{ap} refers to the gain of desired longitudinal driving force, and $F_{x,d}$ refers to the desired longitudinal driving force of full vehicle.

4.1.2 Lateral Control Target Based on Non-linear Model

In the study of stability coordinated control system of vehicle, the proper lateral stability control target is designed, so the vehicle can ensure the driving force, and maintain applicable lateral stability, which is a key technology for vehicle coordinated control system. As a result, the reasonably expected vehicle dynamic model shall be designed, in order to obtain the proper lateral stability control target. When the current vehicle dynamic control target is designed, the linear two-degree of freedom vehicle model is used for designing the control target [1]; although the linear two-degree of freedom vehicle model is featured with concise type and convenient calculation, the model does not consider the change of cornering stiffness of tire, so the dynamic properties of vehicle cannot be accurately reflected. To accurately reflect the motion state of vehicle, and consider the real-time performance of computation, non-linear vehicle model considering the change of cornering stiffness of vehicle was built, with the specific process as follows: Generally, the stability control of the vehicle is only performed in the high-speed running process, so the front wheel steering angle is small, i.e.:

$$\delta \doteq 0 \tag{4.2}$$

The lateral motion equation of vehicle is:

$$mv_x(\dot{\beta} + \gamma) = F_{1,y} + F_{2,y} + F_{3,y} + F_{4,y} \tag{4.3}$$

The yaw motion equation of vehicle is:

$$I_z\dot{\gamma} = l_f(F_{1,y} + F_{2,y}) - l_r(F_{3,y} + F_{4,y}) + M_z \tag{4.4}$$

Because the front wheel of the studied vehicle is turned and the steering angle is small, the side slip angle of tire α_i can be simplified into the Eq. (4.5).

$$\begin{cases} \alpha_1 = \delta - \beta - \dfrac{l_f}{v_x}\gamma \\ \alpha_2 = \delta - \beta - \dfrac{l_f}{v_x}\gamma \\ \alpha_3 = -\beta + \dfrac{l_r}{v_x}\gamma \\ \alpha_4 = -\beta + \dfrac{l_r}{v_x}\gamma \end{cases} \tag{4.5}$$

wherein, the side slip angle β and yaw rate γ can be obtained from the state observer in Chap. 3.

The cornering stiffness of tire will greatly influence the vehicle model. Therefore, to improve the precision of model, the cornering stiffness is subject to linearization treatment at current point, and the lateral force $F_{i,y}$ can be observed in Chap. 3. The accurate cornering stiffness of tire can be evaluated by the observed lateral force $F_{i,y}$ and the simplified side slip angle α_i, as shown in Eq. (4.6).

$$C_i = \frac{F_{i,y}}{\alpha_i} \tag{4.6}$$

The cornering stiffness C_i in Eq. (4.6) is taken as the known number to be substituted into the Eqs. (4.3) and (4.4), the side slip angle β and yaw rate γ are taken as controllable variables, and the direct yaw moment M_z is taken as the external control input and δ as the front wheel steering angle of system. The design state equation is:

$$\dot{x} = Ax + Bu + EM \tag{4.7}$$

wherein,

$$x = \begin{bmatrix} \beta \\ \gamma \end{bmatrix}, E = \begin{bmatrix} 0 \\ e_2 \end{bmatrix} = \begin{bmatrix} 0 \\ \dfrac{1}{I_z} \end{bmatrix}, u = \delta, M = M_z, B = \begin{bmatrix} b_1 \\ b_2 \end{bmatrix} = \begin{bmatrix} \dfrac{C_1 + C_2}{mv_x} \\ \dfrac{l_f(C_1 + C_2)}{I_z} \end{bmatrix}$$

$$A = \begin{bmatrix} a_{11} & a_{12} \\ a_{21} & a_{22} \end{bmatrix} = \begin{bmatrix} \dfrac{-(C_1 + C_2 + C_3 + C_4)}{mv_x} & \dfrac{-l_f(C_1 + C_2) + l_r(C_3 + C_4)}{mv_x^2} - 1 \\ \dfrac{-l_f(C_1 + C_2) + l_r(C_3 + C_4)}{I_z} & \dfrac{-l_f^2(C_1 + C_2) - l_r^2(C_3 + C_4)}{I_z v_x} \end{bmatrix}$$

The non-linear vehicle model in Eq. (4.7) can well reflect the expected response properties of drivers.

4.1.3 Feedforward Control Based on Non-linear Vehicle Model

The conversion Equation of front wheel steering angle δ and direct yaw moment M_z are as follows according to the side slip angle β and yaw rate γ obtained by Eq. (4.7):

$$\dot{\beta} = a_{11}\beta + a_{12}\gamma + b_1\delta \tag{4.8}$$

$$\dot{\gamma} = a_{21}\beta + a_{22}\gamma + b_2\delta_f + e_2 M_z \tag{4.9}$$

To improve the stability of vehicle and the feeling of driver and passengers, the side slip angle shall be minimized, and the control target of design side slip angle is:

$$\beta_d = 0 \tag{4.10}$$

When the vehicle state is balanced, the yaw rate and side slip angle are constant, thus:

$$\dot{\beta} = 0 \tag{4.11}$$

$$\dot{\gamma} = 0 \tag{4.12}$$

At this moment, if the side slip angle of the vehicle can be controlled, the side slip angle meets the control target:

$$\beta = \beta_d \tag{4.13}$$

When Eqs. (4.10)–(4.13) are substituted into Eqs. (4.8) and (4.9):

$$a_{12}\gamma + b_1\delta = 0 \tag{4.14}$$

$$a_{22}\gamma + b_2\delta + e_2 M_z = 0 \tag{4.15}$$

According to Eqs. (4.14) and (4.15), when the vehicle state is balanced, the steady-state value of target yaw rate is:

$$\gamma_0 = -\frac{b_1}{a_{12}}\delta \tag{4.16}$$

When the vehicle is stable, the inputted direct yaw moment is the feedforward moment M_{ff}.

$$M_{ff} = G_{ff}\delta \tag{4.17}$$

wherein, G_{ff} refers to the feedforward gain coefficient.

$$G_{ff} = \frac{a_{22}b_1 - a_{12}b_2}{a_{12}e_2} \tag{4.18}$$

The feedforward control can directly obtain the feedforward yaw moment according to the system input, therefore the system speed of lateral control is improved, and the real-time performance of control system is enhanced.

4.1.4 Design of Expected Vehicle Response Model

When the Eq. (4.17) is substituted into Eq. (4.9), and is simultaneous with the Eq. (4.8):

$$\dot{\gamma} = a_{22}\gamma + \frac{a_{22}b_1}{a_{12}}\delta + a_{21}\beta \tag{4.19}$$

As the control target of side slip angle is 0, and $\dot{\beta} = 0$ is designed in the vehicle driving process. To simplify the control equation, the $\beta = 0$ is substituted into the Eq. (4.19), and the response equation of vehicle expected yaw rate will be:

$$\dot{\gamma}_d = a_{22}\gamma_d + \frac{a_{22}b_1}{a_{12}}\delta \tag{4.20}$$

The Eqs. (4.13) and (4.20) are simultaneous to obtain the expected vehicle response model:

$$\dot{x}_d = A_d x_d + B_d u \tag{4.21}$$

wherein,

$$x_d = \begin{bmatrix} \beta_d \\ \gamma_d \end{bmatrix}, A_d = \begin{bmatrix} 0 & 0 \\ 0 & a_{22} \end{bmatrix}, B_d = \begin{bmatrix} 0 \\ \dfrac{a_{22}b_1}{a_{12}} \end{bmatrix}, u = \delta$$

4.1.5 Optimum Feedback Control Based on LQR

The Sect. 4.1.3 designed the feedforward direct yaw moment, and the real-time performance of system can be enhanced by feedforward control; as the system will suffer from outside interference, it has variable parameters, the stability and robustness of controlled system cannot be guaranteed. For the time change property and non-linear property of vehicle parameters as well as unpredicted external disturbance, it is necessary to design the feedback control to overcome the defect of feedforward control. The side slip angle control target β_d and the yaw rate control target are determined by the front wheel steering angle δ in Eq. (4.21) and the current non-linear vehicle model. The target of feedback control is to follow the side slip angle control target and the yaw rate control target. As shown in Fig. 4.1, the differences of actual states β, γ and control targets β_d, γ_d are inputted into the feedback control module, and then the feedback direct yaw moment M_{fb} is obtained. The stability and robustness of system can be guaranteed by the feedback direct yaw moment. The feedback control error e is defined as the difference of actual state x and control target x_d:

$$e = x - x_d \tag{4.22}$$

According to the Eqs. (4.7) and (4.21):

$$\dot{e} = Ae + EM_{fb} + (A - A_d)x_d + (B - B_d)\delta \tag{4.23}$$

As the last two items in Eq. (4.23) are very small, these two items can be taken as external disturbance and eliminated [2, 3], and the Eq. (4.23) can be simplified as:

$$\dot{e} = Ae + EM_{fb} \tag{4.24}$$

When the error e is reduced, overlarge feedback control amount is not expected, so the feedback direct yaw moment M_{fb} needs to be limited. Therefore, the optimum objective function is:

$$J = \int_0^\infty (e^T Qe + M_{fb}^T RM_{fb}) \tag{4.25}$$

$$Q = \begin{bmatrix} q & 0 \\ 0 & 1-q \end{bmatrix}, q \in [0, 1], R = r_{fb}$$

wherein, Q refers to the weight matrix of state variables, q refers to the weight coefficient of side slip angle β, $1 - q$ refers to the weight coefficient of yaw rate γ, and r_{fb} refers to the weight coefficient of feedback control. The larger the q is, the more important the control over side slip angle will be. The smaller the q is, the more important the control over yaw rate will be. The larger the r_{fb} is, the larger the constraint on feedback yaw moment will be, and the feedback yaw moment control will tend to be reduced. The smaller the r_{fb} is, the smaller the constraint on feedback yaw moment will be, and the feedback yaw moment control will tend to be increased. The Eq. (4.25) is a standard linear quadratic regulator (LQR), so the infinite time optimum control Riccati equation [4] can be established, as shown in Eq. (4.26).

$$PA + A^T P + Q - PER^{-1}E^T P = 0 \tag{4.26}$$

The Riccati equation in Eq. (4.26) is solved, to obtain the optimum feedback control matrix based on LQR:

$$G_{fb} = -R^{-1}E^T P \tag{4.27}$$

The feedback direct yaw moment can be calculated by Eq. (4.28):

$$M_{fb} = G_{fb}^T e \tag{4.28}$$

wherein, G_{fb} refers to the feedback gain coefficient. The feedforward direct yaw moment and feedback direct yaw moment are integrated together to obtain the expected direct yaw moment $M_{z,d}$:

$$M_{z,d} = M_{ff} + M_{fb} \tag{4.29}$$

4.1.6 Adjusting of Weight Coefficient of Lateral Control Target

When the side slip angle of vehicle is very small, the stability of vehicle can be represented by the yaw rate of vehicle relative to the gain of front wheel steering angle. When the side slip angle of vehicle is very large, the same gain will lose the meaning on the representation of the stability of vehicle [5]. If q is used for representing the weigh coefficient of side slip angle in optimum control, when the q is large, the represented vehicle shall follow the β_d; or else, the vehicle shall follow the γ_d. This chapter aims at dynamicly adjusting the weight coefficient q of side slip angle. The dynamic adjusting of q is helpful for the stability of vehicle under different conditions. When the Ref. [5] considered that when the side slip angle is small, the following on automobile yaw rate is the most important; when the side slip angle is large, the following on side slip angle is the most important. According to the reference, the adjusting scheme as shown in Eq. (4.31) could be designed, short for "β method".

$$q = \begin{cases} \dfrac{|\beta|}{\beta_0}, & |\beta| < \beta_0 \\ 1, & |\beta| \geq \beta_0 \end{cases} \tag{4.30}$$

wherein, β_0 is the basis of vehicle stability. According to "β method", when the side slip angle is smaller than β_0, the vehicle is relatively stable, and the side slip angle and yaw rate shall be simultaneously adjusted. Once the side slip angle exceeds the critical value β_0, only the side slip angle is controlled. Such method is too simple for judging the stability of vehicle, which neglects the influence of side slip angle and road surface adhesion coefficient, and cannot obtain better control effect. The Ref. [1] proposed the method for adjusting the dynamic weight coefficient by road surface adhesion coefficient and side slip angle, and the Eq. (4.30) was changed into the Eq. (4.31), short for "$\beta - \mu$ method".

$$q = \begin{cases} \dfrac{|\beta|}{\mu\beta_0}, & |\beta| < \mu\beta_0 \\ 1, & |\beta| \geq \mu\beta_0 \end{cases} \tag{4.31}$$

wherein, β_0 was set as $10°$ in the Ref. [1]. "$\beta - \mu$ method" considers the influence of road surface adhesion coefficient, but the method is only used for judging the stability of vehicle under the quasi-static condition. As the vehicle is the non-linear dynamic system, the vehicle system non-linear properties will be neglected, and the consideration on influence of side slip angle will be lacked when the "$\beta - \mu$ method" is applied. The Ref. [6] considered the influence of side slip angle on weight coefficient q. The dynamic adjusting method used by the reference is as shown in Eq. (4.32), short for "$\beta - \dot{\beta}$ method".

$$
q = \begin{cases} q_0, & |\beta| \le \beta_0 \\ q_0 + \dfrac{|\beta| - \beta_0}{\beta_1 - \beta_0}(1 - q_0), & \beta_0 < |\beta| \le \beta_1 \ \dot{\beta} \le \dot{\beta}_0 \\ 1, & |\beta| > \beta_1 \ \dot{\beta} > \dot{\beta}_0 \end{cases} \tag{4.32}
$$

wherein, when the side slip angle is small, q is fixed; when the side slip angle is large and the change rate of side slip angle velocity is not obvious, q is subject to linear increase along with the side slip angle; when the side slip angle is large or the side slip angle velocity is large, q is 1. The method comprehensively considers the influence of side slip angle and side slip angle velocity, but the consideration on road surface adhesion condition is lacked; when the road surface adhesion condition is poor, the vehicle will be instable even if the side slip angle and side slip angle velocity are small. This dissertation used the $\beta - \dot{\beta}$ phase diagram to judge the side slip state of vehicle, so as to further dynamicly adjust the weight coefficient q. The $\beta - \dot{\beta}$ phase diagram is determined by steering wheel steering angle δ, current vehicle velocity v_x and road surface adhesion coefficient μ [7, 8]. According to the ESP control theory, the $\beta - \dot{\beta}$ phase diagram can reflect the influence of initial motion state on subsequent plane track under stable response condition, and the lateral stability of vehicle can be judged by the $\beta - \dot{\beta}$ phase diagram [7–12]. The $\beta - \dot{\beta}$ phase diagram has stable area and instable area [13, 14]. As shown in Fig. 4.2, the white area is the stable

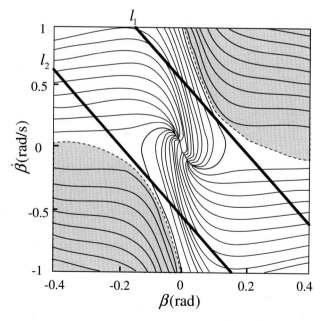

Fig. 4.2 $\beta - \dot{\beta}$ phase plane

area, and the vehicle can restore to the balance state when driving from any point in white area. The shade area in Fig. 4.2 is the instable area, and once the vehicle enters the instable area, the balance state cannot be actively restored, which means that the vehicle instability has occurred.

In Refs. [14–16], academician Guo Konghui, et al. considered that the stable area of the $\beta - \dot{\beta}$ phase diagram can be simplified to the area between two straight lines l_1 and l_2 in Fig. 4.2, as shown in Eq. (4.33).

$$
\begin{cases}
l_1 : E_1\beta + E_2\dot{\beta} = g(\mu) \\
l_2 : E_1\beta + E_2\dot{\beta} = -g(\mu)
\end{cases}
\tag{4.33}
$$

wherein, E_1 refers to the control importance of side slip angle β, E_2 refers to the control importance of side slip angle velocity $\dot{\beta}$, and $g(\mu)$ refers to the increase function relative to μ [14]. As this dissertation mainly discussed the dynamic adjusting of q value, instead of the accurate condition of vehicle instability, so if the vehicle is approximate to the instable state, the q value shall be larger; or else, the q value shall be reduced. The applicable q value can be determined by considering side slip angle, side slip angle velocity and road surface adhesion coefficient. As the $g(\mu)$ is the increasing function relative to μ, according to the study in the Ref. [14], the boundary $g(\mu)$ in stable area is further simplified to the linear function $E_3\mu$ relative to μ.

Accordingly, the q value dynamic adjusting method is as shown in Eq. (4.34), short for "$\beta - \dot{\beta} - \mu$ method".

$$
q =
\begin{cases}
\dfrac{|E_1\beta + E_2\dot{\beta}|}{E_3\mu}, & |E_1\beta + E_2\dot{\beta}| < E_3\mu \\
1, & |E_1\beta + E_2\dot{\beta}| \geq E_3\mu
\end{cases}
\tag{4.34}
$$

wherein, E_1, E_2 and E_3 are the parameters defined by actual vehicle experiments and experience. According to the adjusting method, the instability of vehicle can be caused by overlarge side slip angle or side slip angle velocity. When the side slip angle or side slip angle velocity are not changed, the smaller the road surface adhesion coefficient is, the higher approximation to vehicle stability will be. The adjusting method comprehensively considers the vehicles and road surface, which meets the actual driving experience, and is also consistent with the dynamic theory of vehicle [13, 17, 18]. Compared with the existing q value dynamic adjusting method, the method also considers the influences of side slip angle, side slip rate and road surface adhesion coefficient, the working condition suitability of controller is extended, and the controller is able to ensure the stability of vehicle within the larger range.

4.2 Control Allocation

The previous section defined the desired longitudinal driving force $F_{x,d}$ and direct yaw moment $M_{z,d}$. This section aims at solving the control allocation problem about how to utilize the multi-wheel driving force comprehensive control to reach the demand control target.

4.2.1 Design of Optimum Objective Function for Control Allocation

This section will design the optimum objective function for control allocation. As the distributed electric drive vehicle with multiple power sources is a typical redundancy structure, when part of or all wheels are constrained, optimum driving force control allocation can be performed by comprehensively adjusting the driving force of each wheel, thus ensuring optimum target under certain optimization rule. Each driving motor is subject to coordinated control according to the properties of redundancy configuration of driving unit, which is important to the enhancement of driving capability and safety property of distributed electric drive vehicle. By optimizing the control allocation target, the performance of vehicle dynamic control is improved as follows.

(1) Failure control

For the failure control problem of distributed electric drive vehicle, some scholars proposed the control concept of simultaneously closing failed driving wheels and opposite-side driving wheels. The control method can ensure certain driving capability under the condition of failure of single wheel or double coaxial wheels, with sound robustness and simple control logic [19, 20]. However, the above method does not fully utilize the properties of distributed electric drive vehicle, the failure control problem is not solved on a whole, only part of driving capability of vehicle is guaranteed, and the driving property of vehicle is seriously weakened [21, 22]. At present, there are many studies on the failure control over carrying equipment with redundancy configuration properties of driving unit, which can provide certain concept for the research of topic. The aircraft fault control principle is characterized in that when part of control panel fails, the remained control panels are optimized and combined again, so as to further maintain certain flight capability, and enhance the aircraft survival capability [23, 24]. When the driving part of submariner adopts the redundancy configuration, the fault control is realized by the optimum allocation under the certain optimization rule [25]. The driving source of distributed electric drive vehicle is equipped with redundancy configuration structure, so when part of motors fails or the moment is limited, the driving moment of each driving motor will be subject to coordinated control, so as to reach the desired longitudinal driving force and desired direct yaw moment.

(2) TCS (traction control system)

TCS system can improve the vehicle adhesion capability, enhance the vehicle driving safety, enable the vehicle to smoothly start and drive on the poor road surface [26], and avoid the reduction of lateral stability caused by excessive trackslip in the high-speed running process. However, with ASR, the longitudinal driving capability of driving wheel will be reduced, the vehicle will generate unexpected yaw moment, and lateral stability of automobile will be partly weakened [27]. As the distributed electric drive vehicle is equipped with multiple driving sources, for the vehicles entering TCS working condition, the driving moment of wheels without trackslip are subject to coordinated control, and the longitudinal and lateral dynamic control of vehicle is effectively improved.

This section utilizes the redundancy configuration of driving system, considers the motor properties, introduces the concept of motor utilization rate, and designs the objective function based on optimum allocation while meeting motor constraints. The method is used for performing coordinated control on longitudinal driving property and lateral stability according to different vehicle and motor states.

The main target of driving force control allocation is to meet the desired longitudinal driving force $F_{x,d}$ and direct yaw moment $M_{z,d}$. The driving force control allocation model of distributed electric drive model is as shown in Fig. 4.3.

By controlling the in-wheel driving force of distributed electric drive model, certain longitudinal driving force and direct yaw moment can be generated. By taking the longitudinal driving force and direct yaw moment as control targets, the designed control model can be represented by Eqs. (4.35) and (4.36).

$$F_x = F_{1,x} + F_{2,x} + F_{3,x} + F_{4,x} \tag{4.35}$$

$$M_z = \frac{b_f}{2}(F_{2,x} - F_{1,x}) + \frac{b_r}{2}(F_{4,x} - F_{3,x}) \tag{4.36}$$

The Eqs. (4.35) and (4.36) are simplified into Eq. (4.37).

$$y = Bx \tag{4.37}$$

$$y = [F_x, M_z]^T, \quad x = [F_{1,x}, F_{2,x}, F_{3,x}, F_{4,x}]^T$$

$$B = \begin{bmatrix} 1 & 1 & 1 & 1 \\ -\dfrac{b_f}{2} & \dfrac{b_f}{2} & -\dfrac{b_r}{2} & \dfrac{b_r}{2} \end{bmatrix}$$

wherein, y refers to the controlled variable, x refers to the control input, and B refers to the control matrix. Accordingly, the optimum objective function expressed by Eq. (4.38) is designed.

$$\Omega = \arg \min_{\underline{x} \leq x \leq \overline{x}} \| H_y(Bx - y_d) \|_2 \tag{4.38}$$

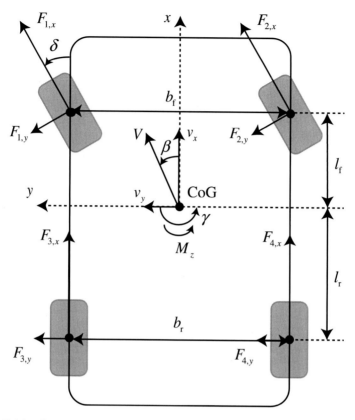

Fig. 4.3 Driving force control allocation model of distributed electric drive vehicle

$$y_d = [F_{x,d}, M_{z,d}]^T, \quad H_y = \text{diag}\{H_F, H_M\}$$

wherein, H_F refers to the weight of longitudinal driving force. The larger the H_F is, the higher the weight meeting the desired longitudinal driving force of driver will be; or else, the lower the weight will be. H_M refers to the weight of direct yaw moment. The larger the H_M is, the higher the weight meeting the desired direct yaw moment of driver will be; or else, the lower the weight will be. The control target y_d is determined by Eqs. (4.1) and (4.29), which reflects the comprehensive demand of drivers. $\underline{x}$ and $\overline{x}$ are the constraints of control input x. The Eq. (4.38) is a typical quadratic programming problem subject to linear constraints [28]. However, as $\text{rank}(B) < 4$, the solution in Eq. (4.38) is not unique, but the solution collection Ω meeting the constraint conditions. When the vehicle velocity is low and the front wheel steering angle is small, the stability of vehicle is good, better dynamics is required, and the main target is to meet the requirement of longitudinal driving force. At this time, the H_F shall be increased, tending to dynamic coordinated control. Along with the increase of vehicle velocity and front wheel steering angle, the vehicle is approximate to instability, better stability is required, and the main target is to meet the requirement of direct yaw moment. At this time, the H_M shall be increased, tending to stability coordinated control. The H_F and H_M are taken as the functions of vehicle velocity and front wheel steering angle, as shown in Eqs. (4.39) and (4.40).

$$H_F = c_1 \exp(-c_2|v_x|) + c_3 \exp(-c_4|\delta|) \tag{4.39}$$

$$H_M = c_5 \exp(c_6|v_x|) + c_7 \exp(c_8|\delta|) \tag{4.40}$$

wherein, $c_i > 0$, $i \in \{1, 2, 3, 4, 5, 6, 7, 8\}$; c_i is obtained by experiments, and is automatically adjusted by weight value. The coordinated control is carried out on dynamics and stability of vehicle. The utilization rate of motor is defined as η, as shown in Eq. (4.41).

$$\eta_i = \frac{F_{i,x}}{F_{i,\max} + \zeta} \tag{4.41}$$

wherein, $F_{i,\max}$, max refers to the maximum in-wheel driving force of distributed electric drive vehicle. To prevent the denominator from generating 0, ζ is a small positive integer. When η becomes larger, the driving moment generated by motor will be approximate to the saturation limit, and the remained driving capability of motor will be smaller. When η becomes smaller, the remained driving capability of motor will be larger. When part of motor fails, part of moment can still be generated; however, the concept of decreasing optimum degree of failure driving source is generally adopted in order to protect the motor [24]. Therefore, the quadratic optimizing scheme is proposed for comprehensively reducing the utilization rate of all driving wheels, and the objective function in Eq. (4.42) is designed.

$$u = \arg \min_{x \in \Omega} \| H_x x \|_2 \tag{4.42}$$

$$H_x = \text{diag}\{\frac{1}{F_{1,\max} + \zeta}, \frac{1}{F_{2,\max} + \zeta}, \frac{1}{F_{3,\max} + \zeta}, \frac{1}{F_{4,\max} + \zeta}\}$$

wherein, H_x refers to the weight matrix of quadratic optimization, which indicates the weights of to-be-distributed driving forces of distributed driving motors. The solution solved by Eq. (4.42) is equivalent to the optimum solution, which considers the requirement of drivers, meets the requirement of constraints of driving wheel moments, and decreases the utilization grade of failure driving wheel. The optimum objective functions (4.38) and (4.42) are comprehensively considered, optimized for simplifying algorithm, and integrated into the objective function (4.43).

$$u = \arg \min_{\underline{x} \leq x \leq \overline{x}} \| H_x x \|_2^2 + \lambda^2 \| H_y (Bx - y_d) \|_2^2 \tag{4.43}$$

As meeting requirement of drivers is very important for the utilization grade of failure wheels, the punish factor λ is set as the larger positive integer ($\lambda \gg 1$).

4.2.2 Establishment of Constraint Conditions of Control Model

The above section designs the optimum objective function of driving force control allocation of distributed electric drive vehicle. This section will comprehensively design the constraint of control input x. The maximum driving moment generated by driving motor is determined by many factors, such as rotate velocity of motor, rated power of motor, peak power, battery voltage, failure mode of motor, overloading running and motor temperature limitation. When the over slip occurs, the vehicle will control the driving motor to completely meet the target driving force established by TCS. In general, the constraints include saturation constraints, slope constraints, road surface adhesion constraints and TCS interference control constraints.

4.2.2.1 Saturation Constraints

The saturation constraints are mainly caused by the motor itself. The constraints include normal load constraints, overload constraints and failure constraints.

(1) Normal load constraints
In theory, the motor generally has the properties of low-speed constant moment and high-speed constant moment. When the rated rotate velocity is approached, quick decrease of driving motor moment will occur. The normal motor load constraints of distributed electric drive vehicle are as shown in Fig. 4.4.

Fig. 4.4 Schematic diagram for motor properties

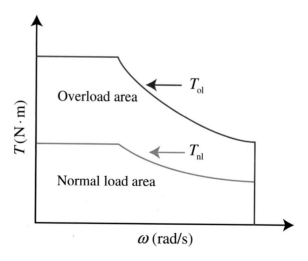

According to Fig. 4.4, the maximum moment T_{nl} generated by the motor will be limited by the current rotate velocity ω under the normal load condition.

$$T_{nl} = g_1(\omega) \tag{4.44}$$

(2) Overload constraints

In the normal load area, the motor can continuously run under normal working condition, and the overheating will not occur. However, when the motor runs in the overload area, the heating amount of motor will be quickly increased, and the working condition will be worse, so it is necessary to limit the driving moment in the overload running process, in order to ensure the safety and prolong the service life of motor [29]. With different temperatures of motor, the limitation on driving moment is also different; in general, under the overload condition, the higher the motor temperature is, the more the limitations of output moment will be, and the lower the maximum driving moment of motor will be. When the upper limit of temperature is reached, the load shall be removed to ensure timely heat radiation and safety of motor. The overload properties of driving motor of distributed electric drive vehicle are as shown in Fig. 4.4. Under the overload condition, the maximum moment T_{ol} generated by the motor will be limited by the current rotate velocity ω and the motor temperature t_c under the overload condition. The constraints can be expressed by Eq. (4.45).

$$T_{ol} = g_2(\omega, t_c) \tag{4.45}$$

(3) Failure constraints

As the distributed electric drive vehicle is a complicated electro-mechanical system with multiple parts, although the previous effective design can minimize the faults, failure and fault cannot be completely avoided after long-time working [30]. It's necessary to design the constraints caused by failures, and improve the vehicle running state under the failure condition. There are many failure conditions of motor, including undercurrent, undervoltage, overheat, overcurrent, overvoltage and insulation failure, which will cause the failure of motor. For the electric vehicle using driving source as single motor, the motor failure usually means that the driving property of vehicle is greatly decreased, and the emergency shutdown occurs. However, for the distributed electric drive vehicle with redundancy configuration of driving source, the failure of single motor will not affect the operation of other motors, and the other driving motors can still be used for ensuring the driving capability of vehicle. The failure constraints of motor moment refer to the limitation on maximum moment generated by motor after the motor failure occurs. After the fault occurs, the maximum moment T_{fc} generated by motor is influenced by the current failure condition, and k is defined as the failure mode of current motor. According to different failure modes, the moment limitation generated by motor is defined as T_{fc}, as shown in Eq. (4.46).

$$T_{fc} = g_3(k) \tag{4.46}$$

The zone bit of overload is defined as τ_{ol}, as shown in Eq. (4.47).

$$\tau_{ol} = \begin{cases} 0, & \text{Normal} \\ 1, & \text{Overload} \end{cases} \tag{4.47}$$

The zone bit of failure is defined as τ_{fc}, as shown in Eq. (4.48).

$$\tau_{fc} = \begin{cases} 0, & \text{Normal} \\ 1, & \text{Fail} \end{cases} \tag{4.48}$$

By comprehensively considering the three saturation constraints, the maximum driving moment generated by motor is T_{max}, as shown in Eq. (4.49).

$$T_{max} = \begin{cases} T_{nl}, & \tau_{fc} = 0 \text{ and } \tau_{ol} = 0 \\ T_{ol}, & \tau_{fc} = 0 \text{ and } \tau_{ol} = 1 \\ T_{fc}, & \tau_{fc} = 1 \end{cases} \tag{4.49}$$

According to Eq. (4.49), the maximum in-wheel driving force generated by driving wheel is F_{max}, as shown in Eq. (4.50).

$$F_{max} = \frac{T_{max}}{R} \tag{4.50}$$

4.2.2.2 Slope Constraints

Different from response properties of engine, the moment response of motor is quick, which is favorable for the dynamic control of distributed electric drive vehicle; however, the over-quick motor response will bring some problems, including frequent impact of driving axle and bearings, causing shorter service life; larger impact of vehicle chassis, causing lower comfortability; damage of motor by frequent moment impact, causing shorter service life. To protect the driving motor and driving axle, and improve the comfortability, the certain slope constraints shall be applied to the driving moment of motor, as shown in Eq. (4.51):

$$r_{min} \leq \dot{T} \leq r_{max} \tag{4.51}$$

wherein, r_{min} and r_{max} are the slope threshold values of motor moment change. The Eq. (4.51) is linearized to obtain the new constraint condition.

$$T(t - \Delta T) + r_{min}\Delta T \leq T \leq T(t - \Delta T) + r_{max}\Delta T] \tag{4.52}$$

wherein, ΔT refers to the time step.

4.2.2.3 Constraints of Road Surface Adhesion Condition

The sizes of longitudinal force and lateral force will be influenced by road surface adhesion, vertical load and tire conditions. Meanwhile, complicated coupling relationship exists between longitudinal force and lateral force. The adhesion ellipse in Fig. 4.5 is used for representing the relationship exists between longitudinal force and lateral force.

When the pure longitudinal trackslip occurs, the longitudinal force will be subject to the following limitations:

$$F_{i,x} \leq \mu_{x,p} F_{i,z} \tag{4.53}$$

wherein, $\mu_{x,p}$ refers to the peak adhesion coefficient of longitudinal force. When the pure lateral side slip occurs, the lateral force will be subject to the following limitation:

$$F_{i,y} \leq \mu_{y,p} F_{i,z} \tag{4.54}$$

wherein, $\mu_{y,p}$ refers to the peak adhesion coefficient of lateral force. For the tire under the combined working condition of lateral side slip and longitudinal slip, the limitation of adhesion ellipse can be expressed by Eq. (4.55) [3].

$$\left(\frac{F_{i,x}}{\mu_{x,p} F_{i,z}}\right)^2 + \left(\frac{F_{i,y}}{\mu_{y,p} F_{i,z}}\right)^2 \leq 1 \tag{4.55}$$

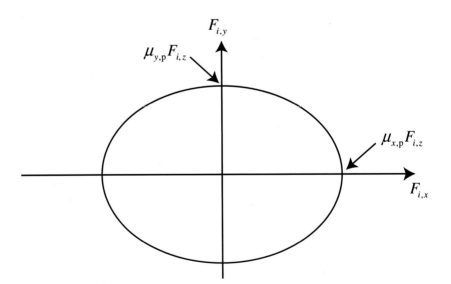

Fig. 4.5 Tire ground adhesion ellipse

Generally, the $\mu_{x,p}$ and $\mu_{y,p}$ can be considered as equivalent [31, 32]. The longitudinal and lateral peak adhesion coefficients are assumed as μ. The friction ellipse can be simplified to friction circle, as shown in Eq. (4.56).

$$\sqrt{F_{i,x}^2 + F_{i,y}^2} \leq \mu F_{i,z} \tag{4.56}$$

The limitation condition of tire friction circle on driving force is obtained from Eq. (4.56), as shown in Eq. (4.57).

$$F_{i,x} \leq \sqrt{\mu^2 F_{i,z}^2 - F_{i,y}^2} \tag{4.57}$$

4.2.2.4 Constraints of TCS

When the vehicle is driven on the road with low adhesion coefficient, the excessive trackslip of tire due to overlarge driving force can occur. At this time, the TCS will issue the moment decrease command to control the trackslip rate, and the driving moment of controlled wheel directly responds to the moment command T_{TCS} of TCS controller [27].

$$T = T_{TCS} \tag{4.58}$$

The constraint of TCS obtained by Eq. (4.58) is an equation constraint. The optimum trackslip rate κ_{opt} of wheel is taken as the control target to design the TCS controller. By reducing the output moment of motor, the actual trackslip rate κ of driving wheel is controlled within the optimal trackslip rate. To realize continuous control and guarantee the robustness, the driving moment of wheel is calculated by the common PID control algorithm. At the time k, the driving moment of TCS system is:

$$T_{TCS}(k) = T(0) - \Delta T_{TCS}(k) \tag{4.59}$$

wherein, k refers to the sampling time, $k = 0, 1, 2, \ldots$, $T(0)$ refers to the driving moment before excessive slip, and $\Delta T_{TCS}(k)$ refers to the to-be-decreased moment value of distributed electric driving wheel at the time k. The input of PID controller is the error signal $e(k)$ of actual trackslip rate and target trackslip rate, as shown in Eq. (4.60).

$$e(k) = \kappa(k) - \kappa_{opt}(k) \tag{4.60}$$

The output is the to-be-decreased moment value $\Delta T_{TCS}(k)$ of excessive trackslip wheel, as shown in Eq. (4.61).

$$\Delta T_{TCS}(k) = K_P e(k) + K_I \sum_{j=0}^{k} e(j) + K_D(e(k) - e(k-1)) \tag{4.61}$$

wherein, K_P refers to the proportional coefficient, K_I refers to the integral coefficient, and K_D refers to the differential coefficient. Therefore, many constraint conditions are summarized into Eq. (4.62).

$$\underline{x} \leq x \leq \overline{x} \tag{4.62}$$

$$\underline{x}_i = \begin{cases} \max\left\{0, \dfrac{T_i(t - \Delta T) + r_{\min}\Delta T}{R}\right\}, & i \notin \Phi_{TCS} \\ \dfrac{T_{i,TCS}}{R}, & i \in \Phi_{TCS} \end{cases}$$

$$\overline{x}_i = \begin{cases} \min\left[\dfrac{T_i(t - \Delta T) + r_{\max}\Delta T}{R}, \dfrac{T_{i,\max}}{R}, \sqrt{\mu^2 F_{i,z}^2 - F_{i,y}^2}\right], & i \notin \Phi_{TCS} \\ \dfrac{T_{i,TCS}}{R}, & i \in \Phi_{TCS} \end{cases}$$

wherein, Φ_{TCS} refers to the collection of wheels with ARS.

4.2.3 Solution of Driving Force Allocation Based on QP

The optimum objective function in Eq. (4.43) can be changed into Eq. (4.63).

$$\lambda^2 \|H_y(Bx - y_d)\|_2^2 + \|H_x x\|_2^2 = \left\| \underbrace{\begin{pmatrix} \lambda H_y B \\ H_x \end{pmatrix}}_{A} x - \underbrace{\begin{pmatrix} \lambda H_y y_d \\ 0 \end{pmatrix}}_{b} \right\|_2^2 \tag{4.63}$$

Therefore, the to-be-solved control allocation problem is converted into a standard linear constraint quadratic programming problem, as shown in Eq. (4.64).

$$\min_{x} \|Ax - b\|_2$$
$$\text{s.t.} \quad Dx \geq U \tag{4.64}$$

wherein,

$$D = \begin{bmatrix} I \\ -I \end{bmatrix}, U = \begin{bmatrix} \underline{x} \\ -\overline{x} \end{bmatrix}$$

The linear constraint quadratic programming (QP) problem can be solved through functional collection, with the calculation steps as follows. The derivation process refers to Ref. [33]. The functional collection is defined as $\mathscr{W}$, including all functional constraints.

(1) Make $x = x^0$ be the feasible initial point, and such point meets all constraint conditions.

(2) For $i = 0, 1, 2, \cdots$ etc., the x^i is sequentially iterated, the optimum correction amount p is found, the inequation constraint in $\mathscr{W}'$ is considered as equation constraint, and other constraints are ignored. Then, the value is further obtained by the Eq. (4.65):

$$\min_{p} \|A(x^i + p) - b\|_2 \qquad\qquad p_i = 0, i \in \mathscr{W} \qquad (4.65)$$

(3) Then, it's necessary to judge if the $x^i + p$ is the feasible point or not; if feasible, the step 4 is performed; or else, the step 5 is performed.

(4) Make $x^{i+1} = x^i + p$, and the Lagrange multiplier λ is calculated.

$$A^{\mathrm{T}}(Ax - b) = D_0^{\mathrm{T}}\lambda \qquad (4.66)$$

wherein, D_0 refers to the initial constraint condition in D. Or else, the constraint corresponding to the minimum value in λ will be removed from $\mathscr{W}$, and the step 2 is performed again

(5) α^i is calculated:

$$\alpha^i = \max\{\alpha \in [0, 1] : C(x^i + \alpha p) \geq U\} \qquad (4.67)$$

Make $x^{i+1} = x^i + p$, the functional constraint on x^{i+1} is added into the $\mathscr{W}$, and the step 2 is performed. The steps (1) to (5) form a complete set of algorithm, using the functional collection method to solve the quadratic programming. The solution is the target in-wheel driving force $F_{i,x}$, and the desired driving moment $T_{i,d}$ distributed to each driving moment can be obtained by Eq. (4.68).

$$T_{i,d} = F_{i,x} R \qquad (4.68)$$

4.3 Motor Property Compensation Control

The steady-state and dynamic response properties of vehicle driving source directly determine the properties of vehicle. For the vehicle with single driving source, even if the driving source has the disadvantage of steady-state error and poor dynamic properties, longitudinal motion condition of vehicle can be corrected by drivers' operation, and the aligning performance of steering system can be guaranteed by the mechanical structure design. For distributed electric drive vehicle with multiple driving sources, the difference of motor will directly influence the effect of dynamic control of vehicle, and if such difference cannot be compensated, the dynamic control of vehicle will be in difficulty, such as poor straight driving stability [6, 34].

Therefore, this section proposes the motor property compensation control strategy based on self-adaptive control method. The strategy specifies the input and output properties of each driving motor, so the steady-state and dynamic properties of each motor are approximately consistent, the desired driving moment of each wheel is accurately executed, and the dynamic property of full vehicle is improved.

4.3.1 Analysis on Motor Response Properties

Although the response speed of motor is higher than that of engine, the motor has the problems on steady-state error and dynamic response properties. The steady-state error of motor are generally generated by the control signal suffering from the influence of outside factors in the transferring process, including static error and proportional error. The dynamic response properties are the inherent properties of motor, and mainly refer to the response model for the motor model which is simplified to first order inertial element. This section will combine the steady-state properties and dynamic properties of electric driving wheels to study the motor property compensation control. The static error refers to the deviation degree of actual output value and theoretical output value of motor at any point, as shown in Fig. 4.6.

In Fig. 4.6, T_d refers to the moment command of motor, T_a refers to the actual output moment of motor, and T_r refers to the theoretical output moment of motor. The theoretical output moment is equal to the inputted moment command, i.e.:

$$T_r = T_d \tag{4.69}$$

However, the actual output moment is:

$$T_a = T_d + T_e \tag{4.70}$$

Fig. 4.6 Schematic diagram for static error of motor

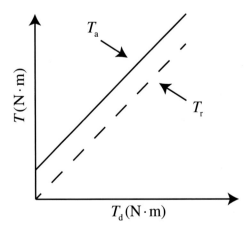

Fig. 4.7 Schematic diagram for proportional error of motor

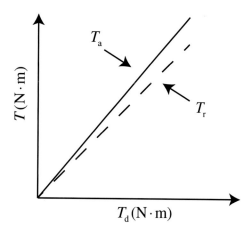

wherein, T_e refers to the static deviation. The proportional error refers to the amplifying relationship between output value and theoretical output value at any point of motor, as shown in Fig. 4.7.

The theoretical output moment shall be equal to the inputted moment command, but the actual output moment will be influenced by the proportional coefficient K, and the actual output is:

$$T_a = K T_d \qquad (4.71)$$

Under most conditions, the static error and proportional error of motor output moment simultaneously exist, and under the steady-state condition, the actual output moment and expected output moment of motor have the relationship as shown in Eq. (4.72).

$$T_a = K T_d + T_e \qquad (4.72)$$

As the distributed driving motor adopts the AC motor in experiment, the control method adopts the direct moment control, and the moment response has no overshoot [35, 36]. Therefore, the input and output property of motor system are expressed by first order inertial transfer function, and the motor model shown in Eq. (4.73) is built.

$$T_a(s) = G_a(s) T_d(s) \qquad (4.73)$$

The motor transfer function model is:

$$G_a(s) = \frac{1}{\tau_a s + 1} \qquad (4.74)$$

wherein, τ_a refers to the response time constant of system. The dynamic error of motor refers to the difference between actual response property and actual response property under the same input condition, as shown in Fig. 4.8.

Fig. 4.8 Schematic diagram
for dynamic response error
of motor

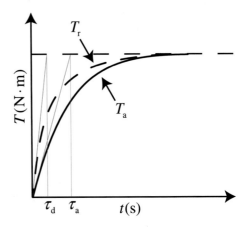

4.3.2 Self-Adaptive Motor Property Control

This section adopts the self-adaptive motor property compensation control based
on model matching [37], so as to compensate the difference of the motor. Firstly,
the standard response properties shall be designed; then, the applicable standard
properties are taken as the expected response properties; furthermore, the difference
between actual response properties and expected response properties is considered
before the reasonable controller is designed, making the actual response properties
be approximate to the expected response properties [38]. This section adopts the
Lyapunov method to design the self-adaptive motor property compensation controller
[39], with the system structure as shown in Fig. 4.9.

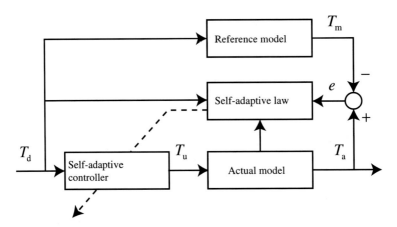

Fig. 4.9 Motor self adaptive compensation control based on Lyapunov method

The reference model is:

$$T_m(s) = G_m(s)T_d(s) \tag{4.75}$$

The transfer function of motor reference model is:

$$G_m(s) = \frac{1}{\tau_d s + 1} \tag{4.76}$$

Solve the Eqs. (4.75) and (4.76) simultaneously to obtain the state equation of reference model:

$$\dot{T}_m = \frac{1}{\tau_d}(T_d - T_m) \tag{4.77}$$

The actual control object model considering steady-state error and dynamic error is:

$$T_a = \frac{K}{\tau_a s + 1}T_u + T_e \tag{4.78}$$

wherein, T_u refers to the outputs of self-adaptive control, which will be sent to the motor. The state equation of actual control object obtained from Eq. (4.78) is:

$$\dot{T}_a = \frac{1}{\tau_a}(KT_u + T_e - T_a) + \dot{T}_e \tag{4.79}$$

The values between the Eq. (4.77) and (4.79) are subtracted as follows:

$$\dot{T}_a - \dot{T}_m = \frac{1}{\tau_a}(KT_u + T_e - T_a) + \dot{T}_e - \frac{1}{\tau_d}(T_d - T_m) \tag{4.80}$$

The deviation e between defined output moment and expected moment is:

$$e = T_a - T_m \tag{4.81}$$

As the static deviation is generally constant, the derivative is 0, i.e.:

$$\dot{T}_e = 0 \tag{4.82}$$

The reference output is:

$$T_m = T_a - e \tag{4.83}$$

When the Eq. (4.83) is substituted into Eq. (4.80):

$$\dot{e} = \frac{1}{\tau_a}(KT_u + T_e - T_a) - \frac{1}{\tau_d}(T_d - T_a + e) \tag{4.84}$$

i.e.,

$$\dot{e} = -\frac{1}{\tau_d}e + \left(\frac{1}{\tau_d} - \frac{1}{\tau_a}\right)T_a + \frac{1}{\tau_a}T_e - \frac{1}{\tau_d}T_d + \frac{K}{\tau_a}T_u \qquad (4.85)$$

or

$$\dot{e} = -\frac{1}{\tau_d}e + P^T Q + \frac{K}{\tau_a}T_u \qquad (4.86)$$

wherein,

$$P = \begin{bmatrix} \frac{1}{\tau_d} & \frac{1}{\tau_a} & \frac{1}{\tau_a} & \frac{1}{\tau_d} \end{bmatrix}^T, \quad Q = [T_a, T_e, T_d]^T$$

The to-be-designed control quantity is:

$$T_u = \theta^T Q \qquad (4.87)$$

When the following deviation e is convergent to 0,

$$e = 0 \qquad (4.88)$$

$$\dot{e} = 0 \qquad (4.89)$$

The control input is:

$$T_u = -\frac{\tau_a}{K}P^T Q \qquad (4.90)$$

The design actual control rate is θ:

$$T_u = \theta^T Q \qquad (4.91)$$

Define:

$$\theta^* = -\frac{\tau_a}{K}P \qquad (4.92)$$

If:

$$T_u = \theta^{*T} Q \qquad (4.93)$$

Then, the following deviation is convergent to 0 by the time constant τ_d. However, as the θ^* is subject to control error, the control rate θ_0 at the initial time can only be:

$$\theta_0 = \theta^* \qquad (4.94)$$

The parameter error is defined as:

$$\phi = \theta - \theta^* \qquad (4.95)$$

Therefore, the Eq. (4.86) can be changed into:

$$\dot{e} = -\frac{1}{\tau_d} e - \frac{K}{\tau_a} \boldsymbol{\theta}^{*\mathrm{T}} \boldsymbol{Q} + \frac{K}{\tau_a} \boldsymbol{\theta}^{\mathrm{T}} \boldsymbol{Q} \tag{4.96}$$

i.e.,

$$\dot{e} = -\frac{1}{\tau_d} e + \frac{K}{\tau_a} \boldsymbol{\phi}^{\mathrm{T}} \boldsymbol{Q} \tag{4.97}$$

The Lyapunov function V is defined, as shown in Eq. (4.98).

$$V = \frac{1}{2}(e^2 + \boldsymbol{\phi}^{\mathrm{T}} \boldsymbol{\Lambda} \boldsymbol{\phi}) \tag{4.98}$$

wherein, $\boldsymbol{\Lambda}$ refers to the weight matrix. The Eq. (4.98) is derived:

$$\dot{V} = e\dot{e} + \boldsymbol{\phi}^{\mathrm{T}} \boldsymbol{\Lambda} \dot{\boldsymbol{\phi}} \tag{4.99}$$

Considering $\dot{\boldsymbol{\theta}}^* = 0$,

$$\dot{V} = e(-\frac{1}{\tau_d} e + \frac{K}{\tau_a} \boldsymbol{\phi}^{\mathrm{T}} \boldsymbol{Q}) + \boldsymbol{\phi}^{\mathrm{T}} \boldsymbol{\Lambda} \dot{\boldsymbol{\theta}} \tag{4.100}$$

When the self-adaptive law in Eq. (4.101) is met,

$$\dot{\boldsymbol{\theta}} = -\frac{K}{\tau_a} e \boldsymbol{\Lambda}^{-1} \boldsymbol{Q} \tag{4.101}$$

The Eq. (4.99) will be:

$$\dot{V} = -\frac{1}{\tau_d} e^2 < 0 \tag{4.102}$$

Therefore, the system meeting the self-adaptive law in Eq. (4.101) is stable, the response properties of each motor moment can be subject to coordinated control by self-adaptive method, and the desired control input is:

$$T_u = (\boldsymbol{\theta}_0 + \int_0^t \dot{\boldsymbol{\theta}} \mathrm{d}t)^{\mathrm{T}} \boldsymbol{Q} \tag{4.103}$$

4.4 Brief Summary

This chapter designed a complete set of coordinated control system of distributed electric drive vehicle, including determination of vehicle dynamic demand target, driving force control allocation, and motor property compensation control. For the determination of vehicle dynamic demand target, the control target of yaw rate and side slip angle was designed on the basis of non-linear vehicle model, and the desired direct yaw moment was designed by the feedforward and feedback joint control mode. In the feedback procedure, the LQR method was adopted for design, so as to improve the robustness of system. This dissertation also proposed the design scheme of using β-phase diagram for the LQR weighting, comprehensively considering the influences of side slip angle, side slip rate and road surface adhesion coefficient, and improving the lateral stability of vehicle. For the driving force control allocation, this dissertation considered many constraints, comprehensively solved the problem of driving force control allocation of distributed electric drive vehicle, and effectively completed the assumed control target even under the conditions of driving wheel failure and excessive slip. To further optimize the utilization condition of each wheel, this dissertation proposed the concept of utilization rate of motor, so as to effectively protect the motor through reasonable objective function. For the motor property compensation control, this dissertation proposed to utilize the model matching method to correct the steady-state and dynamic response properties of multiple motors, in order to make the properties of all motors consistent. Lyapunov method was adopted in the design of self-adaptive control rate, so as to ensure the stability of system.

References

1. Geng C, Mostefai L, Denaï M et al (2009) Direct yaw-moment control of an in-wheel-motored electric vehicle based on body slip angle fuzzy observer. IEEE Trans Ind Electron 56(5):1411–1419
2. Shino M, Miyamoto N, Wang Y et al (2000) Traction control of electric vehicles considering vehicle stability. In: Proceedings of the international workshop on advanced motion control, Japan, Nagoya, Apr 2000, pp 311–316
3. Xiong L, Yu Z, Wang Y et al (2012) Vehicle dynamics control of four in-wheel motor drive electric vehicle using gain scheduling based on tire cornering stiffness estimation. Veh Syst Dyn 50(6):831–846
4. Wu Q, Wang S (2006) Principles of automatic control, No.2 version. Tsinghua University Press, Beijing
5. Shibahata Y, Shimada K, Tomari T (1993) Improvement of vehicle maneuverability by direct yaw moment control. Veh Syst Dyn 22(5–6):465–481
6. Zhang H (2009) Study on torque coordinated control over electric vehicle driven by electric wheel. Jilin University, Changchun
7. Kitahama K (2002) Analysis of vehicles' handling behavior using a phase plane. In: Proceedings of the international symposium on advanced vehicle control, Hiroshima, Japan, July 2002, pp 623–628
8. Inagaki S, Kshiro I, Yamamoto M (1994) Analysis on vehicle stability in critical cornering using phase-plane method. In: Proceedings of the international symposium on advanced vehicle control, Tsukuba, Japan, Oct 1994, pp 287–292
9. van Zanten AT (2000) Bosch ESP systems: 5 years of experience. SAE technical paper: 2000–01–1633
10. Koibuchi K, Yamamoto M, Fukada Y et al (1996) Vehicle stability control in limit cornering by active brake. SAE technical paper 960487

11. Fu H (2008) Study on side slip angle estimation and control strategy for vehicle electronic stability system. Jilin University, Changchun
12. Ko YE, Lee JM (2002) Estimation of the stability region of a vehicle in plane motion using a topological approach. Int J Veh Des 30(3):181–192
13. Pacejka HB (2012) Tire vehicle dynamics, 3rd edn. Elsevier, Oxford
14. Chenchen Z, Xia Q, He L (2011) Study on influence of side slip angle on vehicle stability. J Autom Eng 33(4):277–282
15. Wang D, Guo K, Zong C (2000) Study on theory of vehicle dynamical stability control. J Autom Eng 22(1):7–9
16. Wang D, Guo K, Zong C (1999) Study on simulation of vehicle dynamical stability control. J Autom Technol. J Autom Eng 2:8–10
17. Yu Z (2009) Automobile theory, No.5 version. Tsinghua University Press, Beijing
18. Rajamani R (2012) Vehicle dynamics and control, 2nd edn. Springer, New York
19. Mutoh N, Takahashi Y, Tomita Y (2008) Failsafe drive performance of FRID electric vehicles with the structure driven by the front and rear wheels independently. IEEE Trans Ind Electron 55(6):2306–2315
20. Mutoh N (2009) Front-and-rear-wheel-independent-drive type electric vehicle (FRID EV) with the outstanding driving performance suitable for next-generation adavanced EVs. In: Proceedings of the vehicle power and propulsion conference, Dearborn, Michigan, USA, Sept 2009, pp 1064–1070
21. Kawakami K, Matsugaura S, Onishi M et al (2001) Development of fail-safe technologies of ultra high performance EV "KAZ". In: Proceedings of the international battery, hybrid and fuel cell electric vehicle symposium and exposition, Berlin, Germany, Oct 2001
22. Bo W, Yugong L, Fan J et al (2010) Driving force allocation algorithm of four-wheel independently-driven electric vehicle based on control allocation. J Autom Eng 32(2):128–132
23. Li W, Wei C, Chen Z (2005) New algorithm for limited control direct allocation. J Beijing Univ Aeronaut Astronaut 31(11):1177–1180
24. Harkegard O (2003) Backstepping and control allocation with applications to flight control. LinkOping University, Sweden
25. Omerdic E, Roberts G (2004) Thruster fault diagnosis and accommodation for open-frame underwater vehicles. Control Eng Pract 12(12):1575–1598
26. Hu J, Yin D, Hori Y (2011) Fault-tolerant traction control of electric vehicles. Control Eng Prac 19(2):204–213
27. Chu W, Luo Y, Zhao F et al (2012) Driving force coordinated control of distributed electric drive vehicle. J Autom Eng 34(3):185–189
28. Chen B (2005) Optimal theory and algorithm, No.2 version. Tsinghua University Press, Beijing
29. Zhang Y (2007) Study on power integrated control system for pure electric vehicle. Shanghai Jiaotong University, Shanghai
30. Benbouzid MEH, Diallo D, Zeraoulia M (2007) Advanced fault-tolerant control of induction-motor drives for EV/HEV traction applications: from conventional to modern and intelligent control techniques. IEEE Trans Veh Technol 56(2):519–528
31. Li D (2008) Study on vehicle dynamical integrated control based on optimal allocation. Shanghai Jiaotong University, Shanghai
32. Peng H, Sabahi R, Chen S et al (2011) Integrated vehicle control based on tire force reserve optimization concept. In: Proceedings of the ASME international mechanical engineering congress and exposition, Denver, Colorado, USA, Nov 2011
33. Björck Å (1996) Numerical methods for least squares problems. Society for Industrial and Applied Mathematics, Philadelphia
34. Zhang H, Wang Q, Jin L (2007) Study on the straight-line running stability of the four-wheel independent driving electric vehicles. SAE technical paper: 2007–01–3488
35. Er G, Dou Y (2002) Motion control system. Tsinghua University Press, Beijing
36. Li K (2010) Study on hybrid vehicle dynamical control method based on driving working condition. South China University of Technology, Guang Zhou

37. Oshiumi Y, Shino M, Nagai M (1998) Traction force control of parallel hybrid electric vehicle by using model matching controller. In: Proceedings of the international symposium on advanced vehicle control, Nagoya, Japan, Sept 1998, pp 129–134
38. Li M (2006) Driving torque coordinated control method and hardware in-loop simulation for hybrid vehicle. Tsinghua University, Beijing
39. Sastry S, Bodson M (2011) Adaptive Control: stability, convergence and robustness. Dover Publications, Mineola

Chapter 5
Simulation Verification on State Estimation and Coordinated Control

Abstract Chapters 3 and 4 discussed the state estimation and coordinated control of distributed electric drive vehicle, so as to provide the state estimation method and coordinated control method for verification, and develop the dynamics simulation platform for the distributed electric drive vehicle. The simulation platform can be used for performing repeated simulation under different working conditions, and comparing and analyzing the state estimation and dynamic control effects of motion states of full vehicle under different adhesion conditions, different motion states and different drivers' operation conditions, so as to verify the key development technology. The simulation results indicate that the proposed State Estimation system has higher precision and better real-time property and robustness, and the proposed vehicle coordinated control system comprehensively improves the longitudinal and lateral properties of vehicle under the conditions of drive failure, drive trackslip, etc., so as to ensure the safety of vehicle.

5.1 Development of Simulation Platform

The development of simulation platform shall meet the requirements of high precision, real-time property and repetitiveness. Firstly, only the simulation platform with high precision can be used for developing the vehicle state estimation and coordinated control system; then, the developed algorithm is applied into the actual vehicle, so the real-time property in the simulation process shall be met; finally, to design the reasonable control parameters, the effects of different algorithms under the same working condition shall be repeatedly compared for multiple times, so as to meet the requirement of repetitiveness.

5.1.1 Development of CarSim and Simulink Joint Simulation Platform

CarSim is the special simulation software of vehicle dynamics developed by Mechanical Simulation Corporation, the main founders of which are famous vehicle

© Springer-Verlag Berlin Heidelberg 2016

W. Chu, *State Estimation and Coordinated Control for Distributed
Electric Vehicles*, Springer Theses, DOI 10.1007/978-3-662-48708-2_5

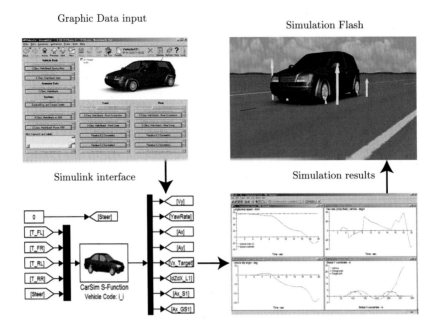

Fig. 5.1 Flow of CarSim and Simulink joint simulation platform

dynamics experts Sayer and Gillespie. CarSim is featured with high calculation precision, good real-time property and repetitiveness, and can accurately reflect the dynamic property of vehicle. At the same time, the programming interface is friendly for parameter setting and vehicle model selection, so as to provide the foundation for the development of novel vehicle dynamic control system. At present, CarSim has been adopted by many famous vehicle manufacturing companies [1–5]. The technology of using the Simulink for developing vehicle controller has been matured. But the vehicle dynamic model developed by Simulink technology is complicated. Therefore, this dissertation developed the dynamic state parameter control and coordinated control simulation platform of distributed electric drive vehicle based on CarSim and Simulink. The simulation process of simulation platform is shown in Fig. 5.1.

Firstly, the data is inputted in the graphic CarSim interface, and the reasonable vehicle dynamic model and simulation scene are set. Secondly, the reasonable control interface and control algorithm are designed in Simulink, and the control interfaces of CarSim and Simulink are butted. Thirdly, the CarSim and Simulink joint simulation is performed to obtain the simulation results. Finally, the simulation cartoons in the simulation process can be obtained, so that the simulation effect will be more visual.

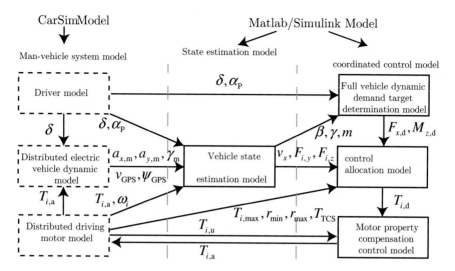

Fig. 5.2 Structure of simulation platform for state estimation and coordinated control

5.1.2 Design of Overall Structure of Simulation Platform

The simulation platform is jointly built by CarSim and Simulink, wherein the man-vehicle model is built by CarSim, and the vehicle state estimation model and coordinated control model are built by Matlab/Simulink. Refer to Fig. 5.2 for the structure of simulation platform.

The man-vehicle model is built by CarSim, including driver model, distributed electric drive vehicle dynamic model and distributed driving motor model. The driver model is used for simulating the activities of driver in the simulation process, i.e. acceleration, braking and steering. The steering operation includes open-loop operation and closed-loop operation. During open-loop steering operation, the driver completes the applicable steering angle of steering wheel according to the instruction. During closed-loop steering operation, the lateral driver model designed by MacAdam is adopted, and the steering is completed according to the expected path provided by simulation scene, and then the expected path is followed [6, 7]. The driver completes the applicable operation according to the designed acceleration pedal and brake pedal, and the closed-loop driving/braking operation is mainly used to complete the driving/braking operation according to the longitudinal expected vehicle velocity provided by the simulation scene. The distributed electric drive vehicle dynamic model is used for simulating the motion response of the simulation vehicle under various working conditions, i.e. the core of the simulation platform. Therefore, this dissertation built the distributed electric drive vehicle dynamic model based on the CarSim.

Table 5.1 Main parameters of simulation platform for distributed electric vehicle

Parameters	Variable (Unit)	Value
Empty mass	m (t)	2.1
Wheelbase	l (m)	2.6
Tread	b (m)	0.622
Tire rolling radius	R (m)	0.3
Yaw moment of inertia	I_{zz} (kg m^2)	2031.4
Height of centroid	h_g (m)	0.54
Height of roll center	h_r (m)	0.45

According to different functions of each part, the vehicle model built by CarSim can be divided into vehicle body system, suspension system, steering system, drive system, braking system and tires. The vehicle body system is used for calculating the longitudinal, lateral and vertical translation responses of full vehicle and the yaw, pitch and side slip rotation responses of vehicle body, and the inputs are suspension force and drag. The suspension system takes the properties of spring and damper into consideration, and the input is mainly the jumping of tire. The input of steering system is the steering angle of steering wheel, and the outputs are deflection angle and lateral force of tire. The drive system mainly refers to the drive shaft from driving motor to hub. The braking system is designed into traditional hydraulic braking. The tire is built by Magic Formula Model [8]. The distributed driving motor model shall reflect the dynamic properties of motor, and can be built by CarSim. Refer to Table 5.1 for the main parameters of simulation platform for distributed electric vehicle.

The vehicle state estimation model is built by the Matlab/Simulink, and is used for observing multiple dynamics parameters of full vehicle, i.e. mass estimation based on high-frequency information extraction, slope estimation based on multi-method fusion, tire vertical force estimation based on multi-information fusion, and vehicle motion state and lateral force estimation based on unscented particle filter. The observed information is used for coordinated control system and other components of vehicle controller. The coordinated control model is built by Matlab/Simulink, and is used for performing coordinated control on driving force of each driving wheel, i.e. determination of vehicle dynamic demand target, driving force control allocation, and motor property compensation control, the three layers of coordinated control to be realized.

5.2 Simulation of State Estimation

The state estimation system of distributed electric drive vehicle includes four parts, i.e. mass estimation based on high-frequency information extraction, slope estimation based on multi-method fusion, tire vertical force estimation based on multi-information fusion, and vehicle motion state and lateral force estimation based on unscented particle filter. Refer to Fig. 2.2 for the relationship of the four parts. During simulation, applicable feature working condition is selected to verify each algorithm according to different to be observed quantities. When the mass and gradient are estimated, the straight driving condition on road with high adhesion coefficient is considered. When the vertical force is estimated, to verify the influence of rolling of vehicle body on estimation, the double motion path working condition and continuous steering condition on road with high adhesion coefficient are considered. When the vehicle motion state and lateral force are estimated by using unscented particle filter, the complexity of vehicle and road is considered, the typical condition is selected, and the straight driving and continuous steering conditions on road with high adhesion coefficient, and the braking, acceleration, double motion path and continuous steering conditions on road with low high adhesion coefficient are considered.

5.2.1 Simulation for Mass Estimation Based on High-Frequency Information Extraction

When the algorithm for mass estimation based on high-frequency information extraction is simulated, the change of mass of full vehicle under different working conditions shall be considered, but the mass shall be maintained fixed in the same driving process. At the same time, the influence of road surface gradient on mass estimation shall be verified. Accordingly, the four simulation working conditions, i.e. normal load flat road, loading flat road, normal load slope road and loading slope road, are respectively designed, as shown in Table 5.2. Table 5.2 shows the actual mass of full vehicle, mass estimation initial value and road condition under various working conditions.

To compare and verify the four simulation working conditions, the same longitudinal driving force control method shall be adopted for different simulation working conditions. Therefore, the open-loop operation of driver is adopted in driving operation. Refer to Fig. 5.3 for the resultant force of in-wheel driving force. As the distributed electric driving wheel is adopted, the resultant force F_x of longitudinal driving

Table 5.2 Simulation working conditions of mass estimation based on high-frequency information extraction

No	Kerb mass (t)	Estimated initial mass (t)	Road conditions
1	2.10	1.80	Flat road
2	2.60	1.50	Flat road
3	2.10	1.80	Slope
4	2.60	1.50	Slope

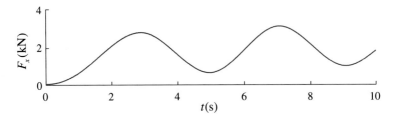

Fig. 5.3 Simulation of mass estimation: working condition of longitudinal driving force

force of each driving wheel is considered as known. Meanwhile, the information of driving acceleration $\dot{v}_x$ is assumed to be obtained in real time, and the standard deviation of driving acceleration noise is $5 \times 10^{-2} \, \text{m/s}^2$. During simulation, the sampling frequency of each variable is about 100 Hz.

In working condition 1, the mass estimation condition of vehicle under normal load condition is analyzed, and the driving on flat road is considered; refer to Fig. 5.4 for the collected longitudinal acceleration information and the estimated mass. In working condition 2, the mass estimation condition of vehicle under loading condition is analyzed, and the driving on flat road is considered; refer to Fig. 5.5 for the collected longitudinal acceleration information and the estimated mass. Figures 5.4 and 5.5 respectively show the driving accelerations and estimated values and true values of mass in the simulation process of mass estimation from top to bottom. In working condition 1, the initial value of mass estimation is 1.8 t, and the true value is 2.1 t; the value is gradually convergent after estimation, after 1.8 s, the estimated value reaches 2.16 t, and the estimation error is 2.9 %. In working condition 2, the initial value of mass estimation is 1.5 t, and the true value is 2.6 t; after 1.4 s, the estimated value reaches 2.66 t, and the estimation error is 2.3 %. Compared with working condition 1 and working condition 2, the results show that the mass estimation

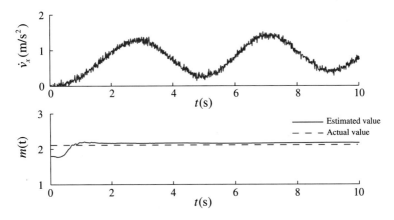

Fig. 5.4 Simulation of mass estimation: flat road, true value of 2.10 t

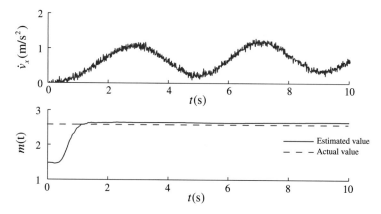

Fig. 5.5 Simulation of mass estimation: flat road, true value of 2.60 t

algorithm is insensitive to the different estimated values of mass, and the values can be quickly convergent to the values near true values under the condition of different true mass values.

The working condition 3 is mainly used for analyzing mass estimation under the normal load condition, the driving on slope road is considered, and the collected road surface gradient, longitudinal acceleration information and the estimated mass are shown in Fig. 5.6. The working condition 4 is mainly used for analyzing the mass estimations under the loading condition, the driving on slope road is considered, and the collected road surface gradient, longitudinal acceleration information and the estimated mass are shown in Fig. 5.7. The Figs. 5.6 and 5.7 respectively show the road surface gradients, driving accelerations, and estimated values and true values of mass in the simulation process of mass estimation from top to bottom, and the road surface gradients are designed as unknown values in the estimation process. In working condition 3, the initial value of mass estimation is 1.8 t, and the true value is 2.1 t; after 1.2 s, the estimated value reaches 2.13 t, and the estimation error is 1.4 %.

In working condition 4, the initial value of mass estimation is 1.5 t, and the true value is 2.6 t; after 1.3 s, the estimated value reaches 2.65 t, and the estimation error is 1.9 %. Compared with working condition 3 and working condition 4, when driving on the slope road, the results show that the mass estimation algorithm is insensitive to the different estimated values of mass, and the values can be quickly convergent to the values near the true values under the condition of different true mass values. After respectively comparing the working condition 1, working condition 2, working condition 3 and working condition 4, it can be seen that the mass estimation algorithm is free from the influence by the road surface gradient, and the mass of full vehicle can be accurately estimated on flat road and slope road. Refer to Table 5.3 for the summary of simulation results of mass estimation. Under various working conditions, the estimation error of mass of full vehicle is less than 2.5 %, and the convergence time is less than 2 s. Under the influences of different gradients and different initial values of mass estimation, the estimated value always can be convergent to near the

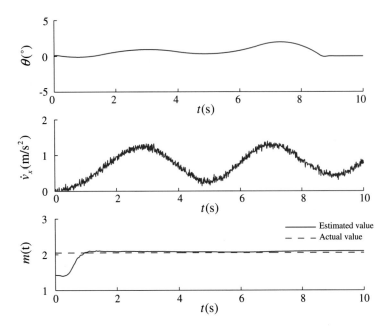

Fig. 5.6 Simulation of mass estimation: slope road, true value of 2.10 t

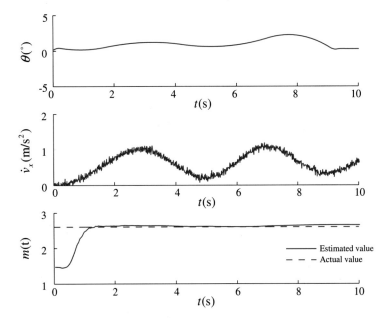

Fig. 5.7 Simulation of mass estimation: slope road, true value of 2.60 t

Table 5.3 Simulation results of mass estimation based on high-frequency information

No	Road conditions	Real values (t)	Estimated values (t)	Error (%)	Convergence time (s)
1	Flat road	2.10	2.16	2.9	1.8
2	Flat road	2.60	2.66	2.3	1.4
3	Slope	2.10	2.13	1.4	1.2
4	Slope	2.60	2.65	1.9	1.3

true value, and the mass estimation algorithm has good robustness on the initial value of mass estimation and the road surface gradient.

5.2.2 Algorithm Simulation for Gradient Estimation Based on Multi-method Fusion

When the algorithm for gradient estimation based on multi-method fusion is simulated, the time varying properties of gradient need to be considered. Accordingly, two different gradient change conditions are designed, including road surface gradient step change and road surface gradient continuous change. Under the two conditions, the vehicle maintains the linear acceleration driving, with the initial acceleration of 20 km/h. Refer to Table 5.4 for the simulation working conditions.

In the simulation process, the resultant longitudinal driving force F_x and the mass of full vehicle m of each driving wheel are designed as known values. Simultaneously, the measured value $a_{x,m}$ of longitudinal acceleration information is obtained by the longitudinal acceleration sensor in real time, and the standard deviation of noise of longitudinal acceleration sensor is 5×10^{-2} m/s^2. In the simulation process, the sampling frequency of each variable is 100 Hz. Figure 5.8 shows the simulation results of gradient estimation under the condition of gradient step change, and shows the longitudinal driving force, longitudinal vehicle velocity, longitudinal acceleration sensor measured value and road surface gradient in the simulation process of gradient estimation (including estimated value by dynamic method, estimated value

Table 5.4 Simulation working conditions of gradient estimation based on multi-method fusion

No	Shape of slope	Longitudinal control	Lateral control	Initial speed (km/h)
1	Step change	Accelerate	Closed loop straight driving	20
2	Continuous changes	Accelerate	Closed loop straight driving	20

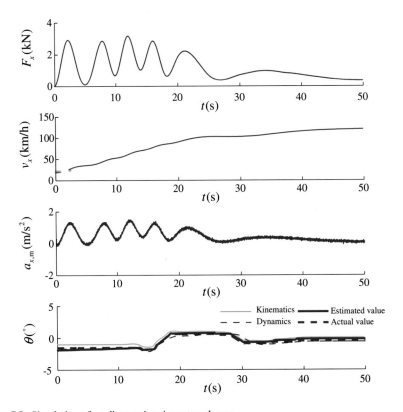

Fig. 5.8 Simulation of gradient estimation: step change

by kinematic method, estimated value by multi-method fusion and true value) from top to bottom.

According to Fig. 5.8, as the kinematic method depends on the precision of longitudinal acceleration sensor, and small static error of longitudinal acceleration sensor will cause larger error of gradient estimation. So static error always exists between the estimated gradient by kinematic method and the true gradient value, especially within the range of 0–13 s. At 16 s, the gradient step change occurs; since the dynamic method adopts the least square method for gradient estimation, it will unavoidably influenced by the previously estimated gradient value, so the value cannot be quickly convergent to near the true value. In order to improve the tracing speed under the condition of gradient step change, and obtain better estimation effects, the gradient estimation method based on multi-method fusion combines the advantages of kinematic method and dynamic method, and removes the static error of gradient estimation.

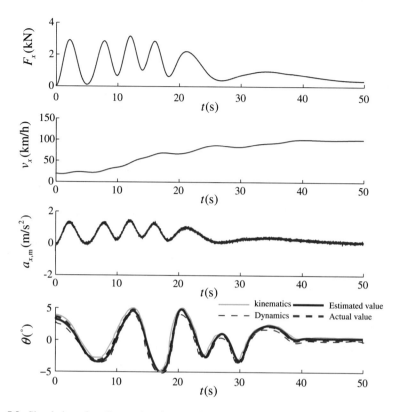

Fig. 5.9 Simulation of gradient estimation: continuous changes

Figure 5.9 shows the simulation results of gradient estimation under the condition of gradient continuous change, and the definition of curve is consistent with that in Fig. 5.8.

According to Fig. 5.9, as the kinematic method depends on the precision of longitudinal acceleration sensor, and small static error of longitudinal acceleration sensor will cause larger error of gradient estimation, so static error always exists between the estimated gradient by kinematic method and the true gradient value. When the gradient change is severe, i.e. at 20–21 s, the estimation results of dynamic method will be greatly influenced by the previously estimated gradient value, so the convergent speed is slow. While at 33–40 s, as the driving force is small, the longitudinal execution on the full vehicle is small, the estimation effect of dynamics effect becomes poorer. Likewise, the gradient estimation method based on multi-method fusion combines the advantages of kinematic method and dynamic method, and removes the static error of gradient estimation, in order to obtain better estimation effects in the continuous gradient change process. Refer to Table 5.5 for the summary of simulation results. Wherein, θ_d refers to the gradient estimation result based on dynamic method, θ_k refers to the gradient estimation result based on kinematic method, and

Table 5.5 Simulation results of gradient estimation on multi method fusion

No	Slope shape	Root mean squared error (°)			Max error (°)		
		θ	θ_d	θ_k	θ	θ_d	θ_k
1	Step change	0.32	0.88	0.88	0.75	2.65	1.97
2	Continuous changes	0.27	0.84	0.86	0.64	1.94	1.87

θ refers to the gradient estimation result based on multi-method fusion. According to Table 5.5, the observation precision of road surface gradient based on multi-method fusion can reach 0.4°; compared with the single use of kinematic method or dynamic method, the precision is obviously increased after the two methods are integrated.

5.2.3 Algorithm Simulation for Vertical Force Estimation Based on Multi-information Fusion

The algorithm for vertical force estimation based on multi-information fusion is simulated on the basis of estimated vehicle mass and road surface gradient. In the simulation and analysis process, firstly, the roll angle and roll rate of vehicle are estimated; then, the longitudinal and lateral acceleration information of vehicle are comprehensively considered; finally, the vertical force of each vehicle is estimated. To verify the estimation effect of algorithm, two simulation working conditions are designed, including uniform velocity double motion path and acceleration continuous steering, as shown in Table 5.6.

In the simulation process, the road surface gradient θ and the mass of full vehicle m are assumed as known values. Simultaneously, the longitudinal acceleration information a_x, lateral acceleration information a_y, and roll rate information $\dot{\phi}$ can be obtained in real time, the standard deviation of longitudinal and lateral acceleration noise is $5 \times 10^{-2}\,\text{m/s}^2$, and the standard deviation of roll rate noise is $0.5°/s$. In the simulation process, the sampling frequency of each variable is 100 Hz. Figure 5.10 shows the vertical force estimation result under the working condition of double motion paths, and respectively shows the estimated value and

Table 5.6 Simulation working condition of vertical force based on multi-information fusion

No	Longitudinal control	Traversal control	Initial speed (km/h)
1	Constant speed	Closed loop, double-lane	60
2	Accelerate	Open loop, continuous steering	50

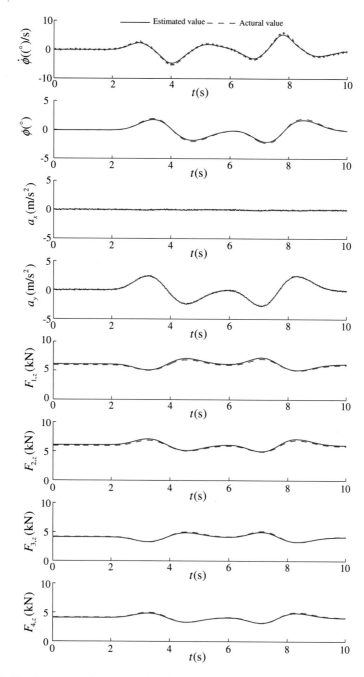

Fig. 5.10 Simulation of vertical force estimation: double lanes

true value of roll rate $\dot{\phi}$, roll angle ϕ, longitudinal acceleration, lateral acceleration and wheel vertical force.

In the double motion path process, the driver drives the vehicle to move with double motion paths at the uniform velocity of 60 km/h; firstly, the roll angle and roll rate are estimated by the method in Sect. 3.3: when the lateral excitation is small (e.g. 2–3 s), the estimated roll angle can well reflect the actual roll angle; when the lateral excitation is large (e.g. near 4 or 8 s), the roll angle is gradually increased. As the non-linear effect of suspension spring is obvious, the error of estimation result is increased but always maintained within the reasonable range, and the maximum estimation error of roll angle is less than 0.2°. Similar to the estimation of roll angle, when the lateral excitation is small, the roll rate is accurate; when the lateral excitation is large, the estimation error of roll rate is increased, and is always maintained within the reasonable range, and the maximum estimation error of roll angle is less than 0.8°/s. The longitudinal and lateral acceleration information is estimated on the basis of roll angle and roll rate, the vertical forces of four wheels are estimated, and the maximum estimation error is 257.5 N, which occurs at the right front wheel at 8.3 s.

Figure 5.11 displays the vertical force estimation results under the working condition of continuous steering motion, and the definition of curves is consistent with that in Fig. 5.10 from top to bottom.

In the continuous steering motion process, the initial vehicle velocity is 50 km/h, the driver drives the vehicle to accelerate at about $0.3 \, \text{m/s}^2$, and the steering wheel angle is controlled in open-loop way. Likewise, when the lateral excitation is small (e.g. near 1 s), the estimated roll angle can well reflect the actual roll angle; when the lateral excitation is large (e.g. near 2 or 4 s), the roll angle is gradually increased. As the non-linear effect of suspension spring is obvious, the error of estimation result is increased but always maintained within the reasonable range, and the maximum estimation error of roll angle is less than 0.4°. Similar to the estimation of roll angle, when the lateral excitation is small, the roll rate is accurate; when the lateral excitation is large, the estimation error of roll rate is increased but always maintained within the reasonable range, and the maximum estimation error of roll angle is less than 1°/s. The longitudinal and lateral acceleration information is estimated on the basis of roll angle and roll rate, the vertical forces of four wheels are estimated, and the maximum estimation error is 312.9 N, which occurs at the right front wheel at 5.7 s.

Refer to Table 5.7 for the summary of estimation results of roll angle, roll rate and vertical force. Table 5.7 shows that the proposed observation method can be used for observing the roll angle and roll rate, and estimation of vertical force can be further completed, with the maximum estimation error of vertical force of about 300 N.

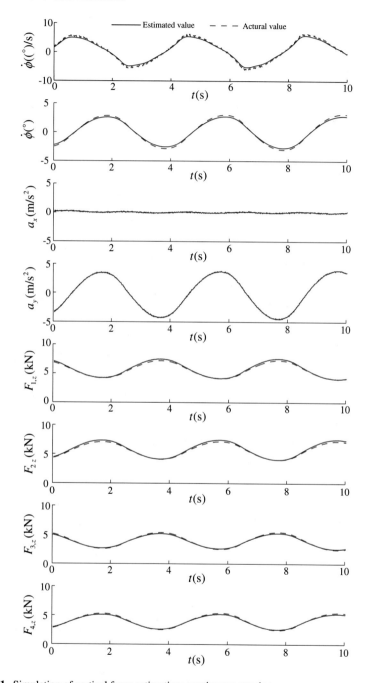

Fig. 5.11 Simulation of vertical force estimation: continuous steering

Table 5.7 Simulation results of vertical force estimation based on multi-information fusion

No	Working conditions	Root mean squared error			Max error		
		$\dot{\phi}$ (°/s)	ϕ (°)	$F_{i,z}$ (N)	$\dot{\phi}$ (°/s)	ϕ (°)	$F_{i,z}$ (N)
1	Double lane change	0.19	7.14×10^{-2}	41.1	0.78	0.20	257.5
2	Continuous steering	0.36	2.26×10^{-2}	101.6	1.03	0.36	312.9

5.2.4 Simulation for Vehicle Motion State and Lateral Force Estimation Based on Unscented Particle Filter

Firstly, the inertial sensor calibration algorithm is simulated and analyzed; then, the different results of vehicle state estimation by different non-linear state estimation methods are compared, and the self-adaptation effect of noise measurement is discussed. Accordingly, the motion state and lateral force of vehicle are jointly estimated by unscented particle filter.

(1) Simulation of inertial sensor calibration algorithm
When the inertial sensor calibration algorithm is simulated and analyzed, two simulation working conditions are designed for the calibration of inertial sensor, including straight driving and continuous steering. Refer to Table 5.8 for the simulation working conditions. It's noticeable that in the simulation process, the initial direction of vehicle velocity always points to due east, so the initial course angle of vehicle is 0.

In the simulation process, the longitudinal acceleration sensor information $a_{x,m}$, lateral acceleration sensor information $a_{y,m}$, yaw rate sensor information γ_m, GPS measured velocity information v_{GPS}, and GPS measured course angle information ψ_{GPS} are assumed to be obtained in real time. The standard deviation of longitudinal acceleration sensor noise is $5 \times 10^{-2}\,\mathrm{m/s^2}$, and the static error $a_{x,b}$ of longitudinal acceleration sensor is $0.2\,\mathrm{m/s^2}$; the standard deviation of lateral acceleration sensor noise is $5 \times 10^{-2}\,\mathrm{m/s^2}$, and the static error $a_{y,b}$ of lateral acceleration sensor is $-0.2\,\mathrm{m/s^2}$; the standard deviation of yaw rate sensor noise is $2.4°/s$, and

Table 5.8 Simulation working conditions of inertial sensor calibration algorithm

No	Longitudinal control	Traversal control	Initial speed (km/h)
1	Constant speed	Closed loop, straight driving	60
2	Constant speed	Open loop, continuous steering	60

the static error γ_b of yaw rate sensor is $7°/s$; the standard deviation of GPS measured velocity noise is 0.1 km/h, and the standard deviation of course angle noise is $6 \times 10^{-3}(°)$. In the simulation process, the GPS measured velocity and course angle sampling frequency is about 1 Hz, and the sampling frequency of other variables is about 100 Hz. Figure 5.12 shows the simulation results of inertial sensor calibration under the working condition of straight driving, and displays the vehicle course angle, GPS measured velocity, measured value by yaw rate sensor, estimated value and true value of static deviation by yaw rate sensor, measured value by longitudinal acceleration sensor, estimated value and true value of static deviation of longitudinal acceleration sensor, measured value by lateral acceleration sensor, and estimated value and true value of static deviation of lateral acceleration sensor from top to bottom. Under the working condition of straight driving, the yaw rate and acceleration sensor are simultaneously calibrated. At the initial time, the initial value of static deviation is 0, and the deviation from the actual value is large. After 1 s, the estimated values of static error of longitudinal and lateral acceleration sensor and yaw rate sensor are quickly convergent to near true values. At this moment, the RMS error of static error $a_{x,b}$ of longitudinal acceleration sensor is about 6×10^{-3} m/s², the RMS error of static error $a_{y,b}$ of lateral acceleration sensor is about 6×10^{-3} m/s², the RMS error of static error γ_b of yaw rate sensor is about $0.32°/s$, and the estimated value of static deviation is very approximate to the true value.

Figure 5.13 shows the simulation results of inertial sensor calibration under the working condition of continuous steering, and the definition of curves is consistent with that in Fig. 5.12 from top to bottom. Under the working condition of continuous steering, on the basis of known static deviation of yaw rate sensor, the longitudinal and lateral acceleration sensor is calibrated. Likewise, at the initial time, the initial value of static deviation is 0. After 1 s, the estimated value of static error of longitudinal and lateral acceleration sensor is quickly convergent to near true value.

Refer to Table 5.9 for the results of inertial sensor calibrated by GPS. When the inertial sensor calibration is simulated, the convergent speed of static deviation of inertial sensor is high–the value near true value can be convergent to near true value within 1 s. The estimated RMS error is smaller than the real error, so the calibration results well reflect the static deviation of inertial sensor.

Table 5.9 Calibration results of inertial sensor

No	Working condition	Convergence time (s)	RMS error of inertial sensor static deviation		
			γ_b (°/s)	$a_{x,b}$ (m/s²)	$a_{y,b}$ (m/s²)
1	Straight driving	1.0	0.32	6.4×10^{-3}	6.1×10^{-3}
2	Continuous steering	1.0	0.32	6.7×10^{-3}	6.2×10^{-3}

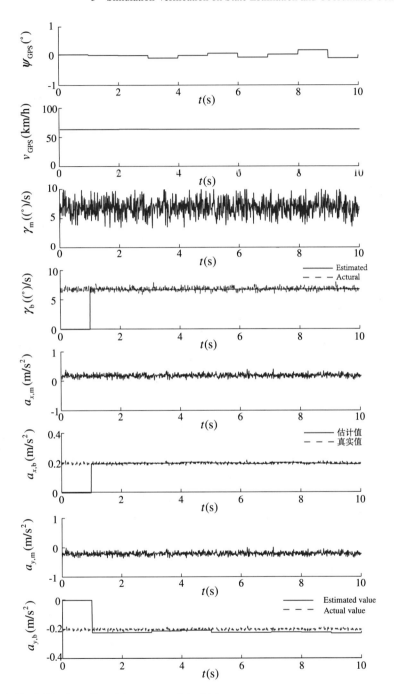

Fig. 5.12 Calibration of inertial sensor: straight driving

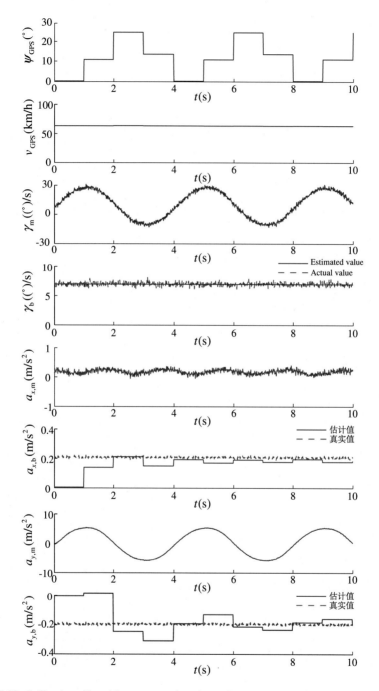

Fig. 5.13 Calibration of inertial sensor: continuous steering

(2) Comparison of non-linear state observers

The Sect. 3.4.1 performed the theoretical analysis on EKF, UKF and UPF. The analysis results showed that as the estimation results of UPF reserves the high order information of non-linear problem, and cancels the Gaussian assumption of state noise, UPF has more advantages when treating the high order non-linear estimation problem.

To better approach the non-linear properties of tire, this dissertation adopts the Magic Formula Model in the estimation process of lateral force, and more accurate state parameter estimation has been realized by the non-linearization of model. As the adopted non-linear state observer mainly influences the estimation precision of lateral force, the influences on the lateral force observation precision by different state observers are compared. To simultaneously compare the observation precision of lateral force by the three different non-linear state estimation methods, the simulation working condition is designed as double motion paths at fixed velocity of 80 km/h; under such condition, the lateral excitation of vehicle is large, and the tires are within the non-linear area, so the observation effects of three state observers can be distinguished.

Figure 5.14 takes the lateral force estimation of right front wheel as example, and shows the results of lateral force observation with UPF, UKF and EKF. When the lateral excitation is large (e.g. near 5 s), the lateral force estimation error of the three state observers are increased because of the non-linear property of tires. Compared with other two methods, since the method with UPF has obvious advantages in treating strong non-linear estimation problem, the observation precision is still high.

Table 5.10 compares the specific observation results of lateral force by three non-linear state estimation methods in Fig. 5.14. The maximum estimation error of lateral force by three state observers occurs in the strong non-linear phase with larger lateral force. The maximum observation error of lateral force by UPF is about 400 N, the maximum observation error is about 1 kN, and the observation error of UKF is about 800 N. Compared with UKF and EKF, the UPF has obvious advantages.

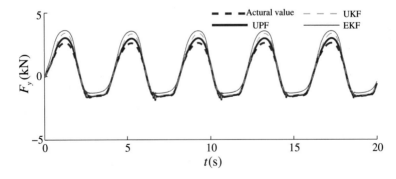

Fig. 5.14 Comparison of non-linear state observers

Table 5.10 Comparison of observation results of non-linear state observers

No	Observation method	RMS error (N)	Max error (N)
1	UPF	129.7	383.6
2	UKF	259.5	767.1
3	EKF	374.2	990.7

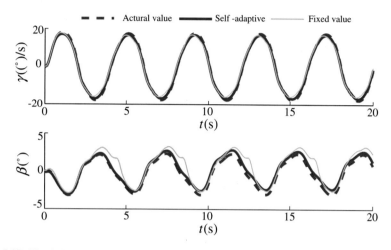

Fig. 5.15 Simulation results of self adaptive adjusting of measure noise

(3) Self-adaptive adjusting of measurement noise

As the self-adaptive adjusting of measurement noise is mainly used to improve the estimation precision of vehicle motion state parameters, the simulation compares the observation results of side slip angle and yaw rate under the conditions of self-adaptive noise adjusting and fixed noise. The selected working condition is designed as continuous steering at the fixed velocity of 80 km/h, and the sampling frequency of each variable is 100 Hz.

Refer to Fig. 5.15 for the simulation results of self-adaptive adjusting of measurement noise. If the fixed noise is adopted, when the side slip angle and yaw rate are large (e.g. 8–9 s), the estimation precision is decreased, the maximum error of side slip angle reaches 2.57°, and the maximum error of yaw rate reaches 3.17°/s. The reason is that the fixed noise excessively depends on the dynamic method, the sharp lateral motion of vehicle is not considered, which will cause severe non-linear effect and decreased precision of dynamic method. In fact, when the side slip angle or yaw rate is large, the reliability of yaw rate sensor and lateral acceleration sensor will be increased, and the estimation effect of kinematic method will be better than that of dynamic method. If the self-adaptation noise is adopted, when the side slip angle and yaw rate are large (e.g. 8–9 s), the reliability of yaw rate sensor and lateral acceleration sensor will be increased, and the observation precision will be

Table 5.11 Simulation results of self adaptive adjusting of measure noise

No	Noise adjustment	RMS error			Max error	
		β (°)	γ (°/s)		β (°)	γ (°/s)
1	Self adaption	0.27	0.77		1.00	1.12
2	Fixed value	0.67	1.59		2.57	3.17

Table 5.12 Simulation working conditions of vehicle longitudinal motion state estimation

No	Adhesion coefficient	Longitudinal control	Initial speed (km/h)
1	0.2	Straight driving, accelerate	20
2	0.2	Straight driving, brake	80

effectively improved. When the self-adaptation noise is applied, the maximum error of side slip angle in the whole estimation process is 1°, and the maximum error of yaw rate is 1.12°/s.

Table 5.11 shows the statistic results of self-adaptive adjusting of measurement noise. The statistic results show that the proposed measurement noise self-adaptation method can effectively improve the observation precision.

(4) Observation of longitudinal motion state

The traditional observation method can well solve the state estimation precision of vehicle in normal motion process, and the observation of longitudinal motion state mainly studies the state estimation problem under the condition of strong driving or strong braking. Accordingly, two simulation working conditions are designed, as shown in Table 5.12, including acceleration on road with low adhesion coefficient, and braking on road with low adhesion coefficient.

Refer to Fig. 5.16 for the simulation results of acceleration on road with low adhesion coefficient. The simulation results compare the estimated value and true value of vehicle velocity, and the in-wheel velocity of four wheels. At 0–6 s, the vehicle acceleration is small, without excessive trackslip; after 6 s, the vehicle acceleration is gradually increased. At 8 s, excessive trackslip occurs to part of wheels; simultaneously, the TCS controller of each driving wheel is adopted, and the slip ratio is reduced by decreasing the driving force of driving wheel. In the acceleration process from 8 to 20 s, excessive trackslip will occur to each wheel; the slip ratio can be maintained within the low rate range by the TCS. In the acceleration process, the trackslip of all wheels even occurs. In the longitudinal large-excitation process, the maximum estimation error of longitudinal velocity of vehicle is 1.11 km/h.

Refer to Fig. 5.17 for the simulation results of braking on road with low adhesion coefficient. The simulation results compare the estimated value and true value of vehicle velocity, and the in-wheel velocity of four wheels. At 0–5 s, the vehicle starts

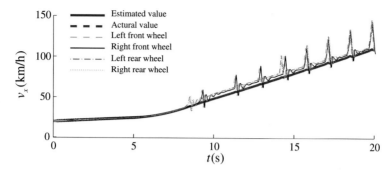

Fig. 5.16   Estimation of longitudinal velocity: acceleration on road with low adhesion coefficient

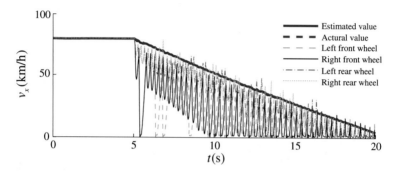

Fig. 5.17   Estimation of longitudinal velocity: braking on road with low adhesion coefficient

the straight driving at uniform velocity of 80 km/h. After 5 s, the vehicle enters the braking condition, and excessive brake locking occurs to part of wheels; simultaneously, the ABS controller of each driving wheel starts the anti-locking brake control, and the slip rate is controlled by decreasing the braking force of driving wheel. In the braking process from 5 to 20 s, excessive slip occurs to each wheel; the slip ratio can be maintained within the low range by ABS. In ABS control process, the brake locking of all wheels even occurs. In the longitudinal large-excitation process, the maximum estimation error of longitudinal velocity of vehicle is 1.08 km/h.

Refer to Table 5.13 for the simulation results of longitudinal motion state estimation of vehicle. According to Table 5.13, the vehicle is in the strong non-linear motion state under the working condition of strong acceleration or strong braking; although part or all of wheels have occurred trackslip or slip, the observation results can effectively reflect the real motion state of vehicle, and the estimation precision is high.

(5) Observation of lateral motion state
This section emphasizes on the state estimation precision under the condition of severe lateral motion, and sets the simulation as fixed-velocity working condition

Table 5.13 Simulation results of longitudinal motion state estimation of vehicle

No	Working condition	RMS error (km/h)	Max error (km/h)
1	Acceleration slip regulation	0.18	1.11
2	Anti lock braking	0.25	1.08

Table 5.14 Simulation working conditions of lateral motion state estimation of vehicle

No	Longitudinal control	Traversal control	Initial speed (km/h)
1	Fixed speed	Closed loop, double lanes	100
2	Fixed speed	Open loop, continuous steering	100

because the above section has observed the longitudinal motion of vehicle. To show the strong non-linearity of tire and road, the road is set as low adhesion coefficient. Accordingly, the two working conditions in Table 5.14 are designed, including double motion paths and continuous steering. In the simulation process, the longitudinal acceleration information a_x, lateral acceleration information a_y, yaw rate information γ, wheel velocity information ω_i, wheel longitudinal driving force information $F_{i,x}$, and steering wheel angle information δ are assumed to be obtained in real time. In the simulation process, the sampling frequency of each variable is 100 Hz.

Figure 5.18 shows the state estimation results under the working condition of double motion paths. As the lateral state parameter in the high velocity motion process is more representative, the longitudinal velocity is maintained near 100 km/h. Under such working condition, the longitudinal velocity v_x, lateral acceleration information v_y, yaw rate γ, and wheel lateral force $F_{i,y}$ can be simultaneously observed.

Figure 5.18 shows the estimated value and true value of longitudinal velocity, lateral velocity, yaw rate, side slip angle and wheel lateral force. The estimation of longitudinal velocity has been discussed in the above section, so this section emphasizes on the observation of yaw rate, side slip angle and wheel lateral force. When the lateral excitation is small, the error of side slip angle will also be small. Along with the increase of lateral excitation, the estimation error of side slip angle is slightly increased. The maximum error of side slip angle occurs at 10 s, which is 0.79°, and the RMS error of side slip angle is 0.23° in the whole estimation process; the maximum error of yaw rate is 0.84°/s, and the RMS error is 0.41°/s; the maximum error of wheel lateral force is 216.9 N, and the RMS error is 60.5 N.

Figure 5.19 shows the state estimation results under the working condition of continuous steering. As the lateral state parameter in the high velocity motion process is more representative, the longitudinal velocity is maintained near 100 km/h. Under such working condition, the longitudinal velocity v_x, lateral acceleration information v_y, yaw rate γ, and wheel lateral force $F_{i,y}$ can be simultaneously observed. The curves are consistent with those in Fig. 5.18. Similiar to the working condition

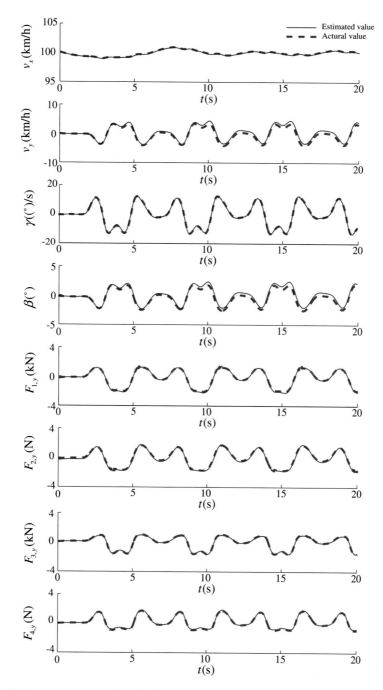

Fig. 5.18 Lateral state estimation: double lanes

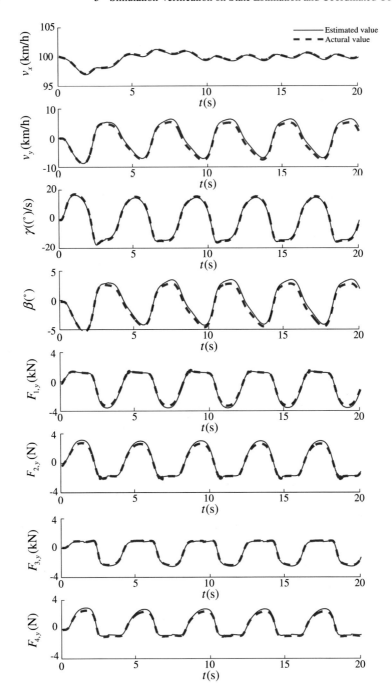

Fig. 5.19 Lateral state estimation: continuous steering

Table 5.15 Simulation results of lateral motion state estimation of vehicle

No	Working condition	RMS error				Max error		
		β (°)	γ (°/s)	$F_{i,y}$ (N)		β (°)	γ (°/s)	$F_{i,y}$ (N)
1	Double lanes	0.23	0.41	60.5		0.79	0.84	216.9
2	Continuous steering	0.28	0.73	157.6		1.05	1.10	423.6

of double lane change, in the continuous steering process, when the lateral excitation is small, the estimation error of side slip angle will also be small; when the lateral excitation is large, the estimation error is slightly increased. The maximum error of side slip angle occurs at 12 s, which is $1.05°$, and the RMS error of side slip angle is $0.28°$ in the whole estimation process; the maximum error of yaw rate is $1.10°/s$, and the RMS error is $0.73°/s$; the maximum error of wheel lateral force is 423.6 N, and the RMS error is 157.6 N.

Refer to Table 5.15 for the simulation results of lateral motion state estimation of vehicle. According to Table 5.15, the vehicle is in the strong non-linear motion state under the working conditions of double lane change and continuous steering, the observation precision of side slip angle, yaw rate and tire lateral force are high, and the observation results can effectively reflect the real motion state of vehicle.

5.3 Simulation of Coordinated Control Algorithm

The coordinated control in this dissertation includes determination of vehicle dynamic demand target, driving force control allocation, and motor property compensation control; the demand target of full vehicle is set, the driving force of each wheel is subject to control allocation according to the control target, the force set for each driving wheel is subject to self-adaptive control, and then the expected moment is followed; for the design of control system, the design flow process from top to bottom is adopted; for the simulation verification, the sequence from bottom to top is adopted. Firstly, the motor property compensation control algorithm is simulated and verified; then, the driving force control allocation algorithm is simulated and verified; finally, the full-vehicle dynamic demand target determination algorithm is simulated and verified. In the simulation process, according to different to-be-verified algorithms, the application feature working conditions are selected, to respectively verify. When the motor property compensation control algorithm is simulated, the working condition of driving on road with high adhesion coefficient is taken into consideration. When the driving force control allocation algorithm is simulated, the high-velocity simulation and low-velocity simulation on road with high adhesion coefficient and bidirectional driving are performed, so as to compare the control conditions of vehicles under normal driving, failure, trackslip, etc. When the full-vehicle dynamic demand target determination algorithm is simulated, the influence of different control algorithms on the stability of full vehicle is compared, and the double motion path simulation and continuous steering simulation are respectively performed on the road with low adhesion coefficient.

5.3.1 Simulation of Motor Property Compensation Control Algorithm

Firstly, the single motor driving moment property compensation control effects under different input conditions are verified. When the motor properties are considered, it is necessary to convert the step control command of controller into expected moment with first order delay property, and then follows the expected moment. When the single motor driving moment self-adaptive control is performed, two different simulation working conditions of step input and continuous input of moment command are observed, as shown in Table 5.16.

Figure 5.20 shows the simulation results of single wheel motor property compensation control under the condition of step input of moment command. Figure 5.20 compares the control command, expected output moment, compensation control output moment and non-compensation output moment from top to bottom. At 0–2 s, the control moment command of vehicle is 0; however, because of static deviation of motor moment, when the compensation control is not performed, the outputted moment command is about 7 N, but not 0. In the compensation control process, the output moment of motor always follows the expected value. At 2 s, step change occurs in motor moment command. If the compensation control is not performed, the dynamic response of output moment of motor is not consistent with the expected value, and the final steady-state value has larger deviation from the actual command. If the compensation control is performed, the dynamic response of output moment of motor and the final steady-state value always can follow the expected output moment. At 6 s, the step of motor moment command occurs again, and the motor output moment after property compensation control can well follow the expected moment.

Figure 5.21 shows the simulation results of single wheel motor property compensation control under the condition of continuous change input of moment command.

Table 5.16 Simulation working conditions of motor property compensation control: single wheel simulation

No	Input torque command	Error exist
1	Step input	Static error, ratio error, dynamic error
2	Continuous input	Static error, ratio error, dynamic error

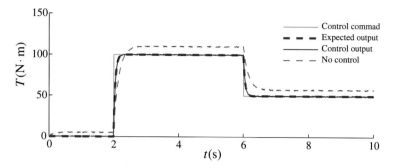

Fig. 5.20 Motor property compensation control (single wheel): step input

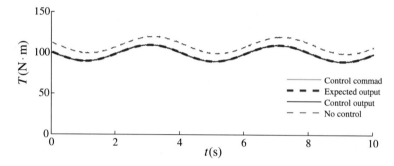

Fig. 5.21 Motor property compensation control (single wheel): continuous input

Figure 5.21 compares the control command, expected output moment, compensation control output moment and non-compensation output moment from top to bottom. In the whole process, the input moment command is always subject to the sine change. When the motor property compensation control is performed, the output moment always can follow the expected output moment. If the motor property compensation control is not performed, static deviation always will exist between the output moment and the expected moment. According to Figs. 5.20 and 5.21, the proposed property compensation control method can improve the moment response properties, and the output moment is controlled to well follow the expected moment.

On the basis of completed single wheel motor property compensation control, the straight driving capabilities of vehicle under the conditions with or without property compensation control are compared. Refer to Table 5.17 for the simulation working conditions. In the simulation process, the steering wheel angle is 0.

Figure 5.22 shows the simulation of driving property of center of mass of vehicle under the condition of uniform velocity. The Figure compares the expected track and actual track of vehicle. When the steering wheel angle is maintained as 0, the expected track is the straight line along the centerline of road. However, by considering the moment error of distributed driving motor, if the property compensation control is

Table 5.17 Simulation working conditions of motor property compensation control: simulation of full vehicle

No	Control method	Error exist	Initial speed (km/h)
1	Constant speed	Static error, ratio error, dynamic error	60
2	Straight acceleration	Static error, ratio error, dynamic error	50

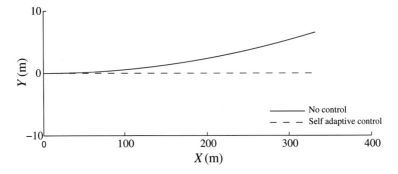

Fig. 5.22 Straight driving property: uniform velocity

not performed, the moments at the two sides of vehicle are different, then the straight driving capability becomes poorer, and the larger lateral displacement occurs. At the end of uniform velocity working condition, the lateral displacement of vehicle will deviate 6.6 m from the preset track. After the motor property compensation control is performed, the straight driving capability becomes better, and the larger lateral displacement does not occur at the end of simulation working condition.

Figure 5.23 shows the simulation of driving property of center of mass of vehicle under the condition of acceleration. The Figure compares the expected track and actual track of vehicle. Similar to the uniform velocity working condition, the expected track is the straight line along the centerline of road. However, if the property compensation control is not performed, the straight driving capability becomes poorer, and at the end of uniform velocity working condition, the lateral displacement of vehicle deviates 8.6 m from the preset track. After the motor property compensation control is performed, the straight driving capability becomes better, and the larger lateral displacement does not occur at the end of simulation working condition.

Therefore, the proposed motor property compensation control algorithm can compensate the steady-state and dynamic response property error of motor, and improve the straight driving capability of vehicle.

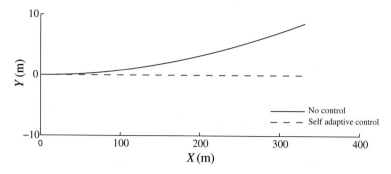

Fig. 5.23 Straight driving property: acceleration

5.3.2 Simulation of Driving Force Control Allocation Algorithm

The simulation of driving force control allocation algorithm mainly considers the conditions of drive trackslip and drive failure in the process of driving force control allocation. To comprehensively consider the dynamic properties and stability, the performance of vehicle under the three working conditions of the coordinated control allocation, non-coordinated control and normal driving (without drive failure and drive trackslip) are compared. Under various working conditions, the open-loop operation by steering wheel angle is adopted, and the steering wheel angle is 0.

In the low velocity phase, the driving force control allocation under the condition of drive failure and drive trackslip is emphasized; according to Sect. 4.2.1, the desired longitudinal driving force is firstly met. Under the working condition of low-velocity trackslip, the vehicle starts and accelerates from the idle state on the bidirectional road. Under the condition of low-velocity failure, the vehicle starts on the road with high adhesion coefficient, partial failure occurs to the right front wheel 1 s after starting, and to left rear wheel 3 s after starting. In the high-velocity phase, the driving force control allocation under the conditions of drive failure and drive trackslip is emphasized; according to Sect. 4.2.1, the desired direct yaw moment is firstly met. Under the condition of high-velocity trackslip, the right wheels always run on the road with low adhesion coefficient, and the left wheels always run on the road with high adhesion coefficient. Under the condition of high-velocity failure, partial failure occurs to the right front wheel 1 s after the vehicle velocity exceeds 80 km/h.

Figure 5.24 shows the dynamic coordinated control results under the working condition of low-velocity starting failure. The Figure compares the vehicle running states under non-failure, non-coordinated control and coordinated control conditions, and sequentially compares the longitudinal driving force, direct yaw moment, longitudinal velocity, motion track, and driving force of each wheel under these three conditions from top to bottom. In the simulation process, the vehicle starts and accelerates from the idle state, the desired longitudinal driving force is stably increased, and the desired direct yaw moment is 0. Firstly, the working condition of

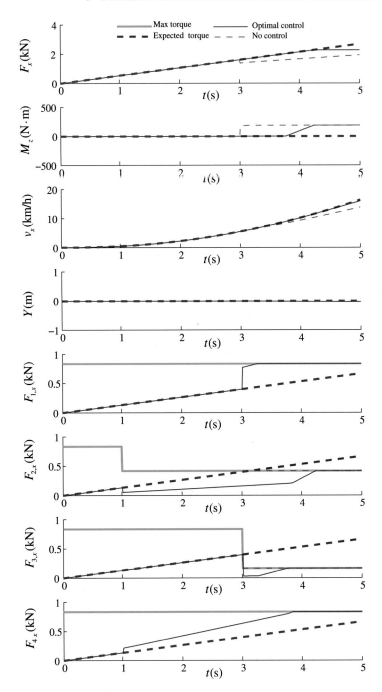

Fig. 5.24 Simulation of dynamic coordinated control: fail in low velocity start

low-velocity starting failure without coordinated control is analyzed. Partial failure occurs to the right front wheel 1 s after starting, and the generated maximum driving force is decreased to 50% of original value, but it is still higher than the expected moment value. Therefore, at 1–3 s, if the coordinated control is not performed, partial failure of right front motor does not influence the longitudinal drive of full vehicle, and the desired moment is still executed. Partial failure occurs to the left rear wheel 3 s after starting, and the generated maximum driving force is decreased to 20% of original value. Simultaneously, the expected moment desired is gradually increased, and the right front motor and left rear motor cannot execute the expected driving force after 3 s; if the coordinated control is not performed, the expected driving force cannot be executed, and the dynamic performance of full vehicle will be greatly influenced; after 5 s, the longitudinal velocity of vehicle is 13.8 km/h. Then, the working condition of coordinated control is analyzed, the obvious advantage is not revealed before 3 s, while after 3 s, as the two wheels fail, the coordinated control system can comprehensively utilize the driving forces of all wheels to meet the requirements of full vehicle, and the driving forces of the other two driving wheels are coordinated and increased. Until before 4.2 s, the expected longitudinal driving force can still be followed. After 4.2 s, the driving forces of all wheels have been saturated; although the driving force is slightly less than the expected longitudinal driving force, it is the optimum driving force under such condition. After 5 s, the longitudinal velocity of vehicle is 15.9 km/h. Under the condition of no failure in low-velocity starting of vehicle, the longitudinal velocity of vehicle is 16.3 km/h after 5 s. According to the comparison results, the coordinated control algorithm based on dynamic property can effectively improve the starting and acceleration capability in the low-velocity starting phase.

Figure 5.25 shows the dynamic coordinated control results under the working condition of low-velocity starting and trackslip. The Figure compares the vehicle running states under non-trackslip, non-coordinated control and coordinated control conditions, and the definition of curves is consistent with that in Fig. 5.24. In the simulation process, the vehicle starts and accelerates from the idle state, the desired longitudinal driving force is stably increased, and the desired direct yaw moment is 0. Excessive trackslip occurs to the right front wheel after 3.4 s, when the TCS is adopted to decrease the driving force command of right front wheel, for the wheel cannot continue to execute the expected driving force. When the coordinated control is not performed, the expected driving force cannot be executed, and the dynamic performance of full vehicle will be greatly influenced. After 5 s, the longitudinal velocity of vehicle is 15.3 km/h. If the coordinated control is performed, the driving forces of the other three wheels can be coordinated and increased to ensure to follow the expected longitudinal driving force; the driving force of right rear wheel is firstly increased, so as to avoid generating overlarge direct yaw moment. After 5 s, the longitudinal velocity of vehicle is 16.2 km/h. Under the condition of no trackslip in vehicle low-velocity starting, the longitudinal velocity of vehicle is 16.3 km/h after 5 s. According to the comparison results, under the condition of trackslip, the coordinated control algorithm based on dynamic property can effectively improve the starting and acceleration capability of vehicle in the low-velocity starting phase.

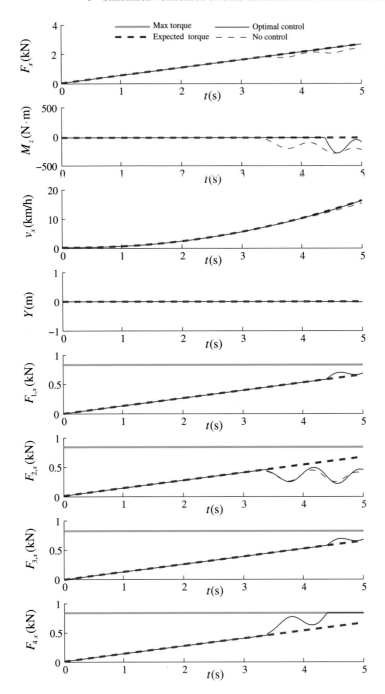

Fig. 5.25 Simulation of dynamic coordinated control: trackslip in low velocity start

Figure 5.26 shows the coordinated dynamic stability control results under the working condition of high-velocity motion. The Figure compares the vehicle running states under non-failure, non-coordinated control and coordinated control conditions, and the definition of curves is consistent with that in Fig. 5.24. In the simulation process, Partial failure occurs to the right front wheel 1 s after the vehicle velocity exceeds 80 km/h, and the wheel cannot continue to execute the expected driving force. When the coordinated control is not performed, as the driving force at right side is suddenly decreased, the unexpected direct yaw moment will be generated, and the lateral stability of full vehicle will be greatly influenced. The lateral displacement of vehicle is 8.6 m 5 s after acceleration, which is dangerous for the vehicle driving at high velocity. If the coordinated control is performed, the unexpected direct yaw moment can be avoided by coordinating and controlling the driving forces of the other three wheels. After 5 s, the lateral displacement of vehicle is about 0.7 m. It can be seen by comparing the lateral displacement of that vehicle, after the coordinated control is performed, the safety of full vehicle is guaranteed.

Figure 5.27 shows the coordinated dynamic stability control results under the working condition of high-velocity motion trackslip. The Figure compares the vehicle running states under non-trackslip, non-coordinated control and coordinated control conditions, and the definition of curves is consistent with that in Fig. 5.24. In the simulation process, the right wheels always run on the road with low adhesion coefficient, and the left wheels always run on the road with high adhesion coefficient. Excessive trackslip occurs to the right front wheel after 0.3 s, when the TCS is adopted to decrease the driving force command of right front wheel, for the wheel cannot continue to execute the expected driving force. When the coordinated control is not performed, as the driving force at right side is suddenly decreased, the unexpected direct yaw moment will be generated, and the lateral stability of full vehicle will be greatly influenced. After 5 s, the lateral displacement of vehicle is 10.5 m, which is dangerous for the vehicle driving at high velocity. If the coordinated control is performed, the unexpected direct yaw moment can be avoided by coordinating and controlling the driving forces of the other three wheels. After 5 s, the lateral displacement of vehicle is about 1.3 m. It can be seen by comparing the lateral displacement of vehicle that, after the coordinated control is performed, the safety of full vehicle is guaranteed.

Refer to Table 5.18 for all simulation results. It's noticeable that as the lateral displacement less than 0.1 is of less significance, it can be regarded as 0. According to the simulation results, the coordinated control algorithm based on dynamic property can effectively improve the starting and acceleration capability of vehicle under the conditions of failure (working condition 2) and trackslip (working condition 4) in the low-velocity starting phase. After 5 s of starting, the vehicle longitudinal velocity subject to coordinated control algorithm is almost equal to that under normal running working condition (working condition 1), and the starting and acceleration performance are obviously superior to those without coordinated control (working condition 3 and working condition 5). Although certain unexpected yaw moment is generated (target direct yaw moment of 0), as the vehicle velocity is low, large lateral displacement does not occur, and the relative stability is maintained. In the

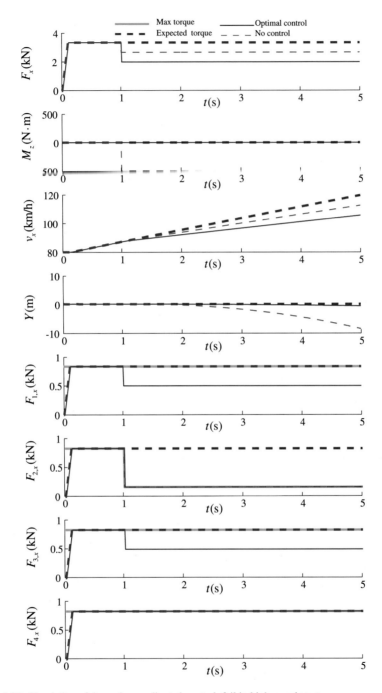

Fig. 5.26 Simulation of dynamic coordinated control: fail in high speed start

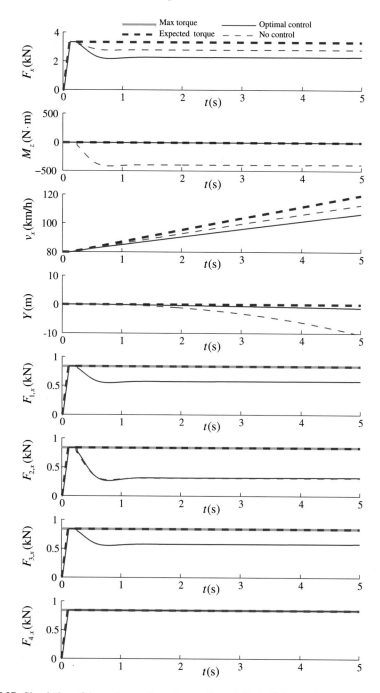

Fig. 5.27 Simulation of dynamic coordinated control: trackslip in high speed start

Table 5.18 Simulation working conditions of driving force control allocation

No	Working condition	Failure	Slip	Control coordination	End speed (km/h)	Lateral disp.(m)
1	Start at low speed	No failure	No slip	Average distribution	16.3	0
2	Start at low speed	Fail	No slip	Dynamic coordination	15.9	0
3	Start at low speed	Fail	No slip	No coordination	13.8	0
4	Start at low speed	No failure	Slip	Dynamic coordination	16.2	0
5	Start at low speed	No failure	Slip	No coordination	15.3	0
6	Start at high speed	No failure	No slip	Average allocation	119.5	0
7	Start at high speed	Fail	No slip	Steady coordination	105.6	−0.7
8	Start at high speed	Fail	No slip	No coordination	112.5	−8.6
9	Start at high speed	No failure	Slip	Steady coordination	106.6	−1.3
10	Start at high speed	No failure	Slip	No coordination	112.9	−10.5

high-velocity motion phase, the lateral stability of vehicle can be effectively guaranteed under the conditions of failure (working condition 7) and trackslip (working condition 9) by the coordinated control algorithm based on stability. At 5 s after acceleration, the lateral displacement of vehicle subject to coordinated control algorithm is only about 1 m; as the vehicle runs for 120 m within 5 s, the driver is completely capable of ensuring the stable driving of vehicle in such period. If the coordinated control is not performed (working condition 8 and working condition 10), at 5 s after acceleration, the generated lateral displacement will have reached about 10 m, which is very dangerous in the high-velocity running process. In general, the coordinated control algorithm can effectively improve the starting and acceleration capability of vehicle in the low-velocity phase, and ensure the stability of vehicle in the high-velocity phase.

5.3.3 Simulation of Determination Algorithm for Vehicle Dynamic Target

The simulation of determination algorithm for full-vehicle dynamic demand target is mainly used to verify if the proposed algorithm can effectively control the side

Table 5.19 Simulation working conditions of vehicle dynamic demand target

No	Adhesion coefficient	Long. Control	Traversal control	Initial speed (km/h)
1	0.25	Constant speed	Closed loop, double-lane	60
2	0.15	Constant speed	Open, continuous steering	100

Table 5.20 Scheme for lateral stability control target adjusting

No	Description	Abbreviation
1	Only β	β method
2	Only β and μ	$\beta - \mu$ method
3	Only β and $\dot{\beta}$	$\beta - \dot{\beta}$ method
4	Consider β $\dot{\beta}$ and μ	$\beta - \dot{\beta} - \mu$ method
5	No control	No control

slip angle and yaw rate of vehicle, so as to further ensure the safety of vehicle. In the simulation process, the road surface adhesion coefficient is low, and the vehicle is subject to severe lateral motion. Refer to Table 5.19 for the designed simulation working conditions.

To compare the control effects of different control demand targets, except for the working condition without control, this dissertation also designed other three weighting coefficient adjusting methods for main lateral control targets (refer to Sect. 4.1.6), and summarized five control schemes, as shown in Table 5.20.

Figure 5.28 shows the lateral control effects under the working condition of double lane change. In the simulation process, the longitudinal velocity of vehicle is about 60 km/h, and the road pavement coefficient is 0.25. The Figure also compares the five control schemes shown in Table 5.20. Figure 5.28 sequentially compares the side slip angle, yaw rate, motion track, and $\beta - \gamma$ phase diagram under five control schemes from top to bottom. If direct yaw moment control is not performed, after 6 s, the side slip angle under the working condition without control will have entered the instable area. After direct yaw moment control is performed, the several schemes can control the vehicle within the stable area, and only the control effects are different. The "$\beta - \dot{\beta} - \mu$ method" can decrease the maximum side slip angle to 0.52°, while the maximum yaw rate is 8.11°/s, and the lateral stability of vehicle is good. When the "β method" is adopted, the maximum side slip angle reaches 4.05°, and the maximum yaw rate is 8.11°/s; although the vehicle is not instable, it has approached to instability, so the control effect is poor. The effect of "$\beta - \mu$ method" and the "$\beta - \dot{\beta}$ method" are between the above two methods.

Figure 5.29 shows the lateral control effects under the working condition of continuous steering. In the simulation process, the longitudinal velocity of vehicle is about 100 km/h, and the road pavement coefficient is 0.15. The Figure also compares

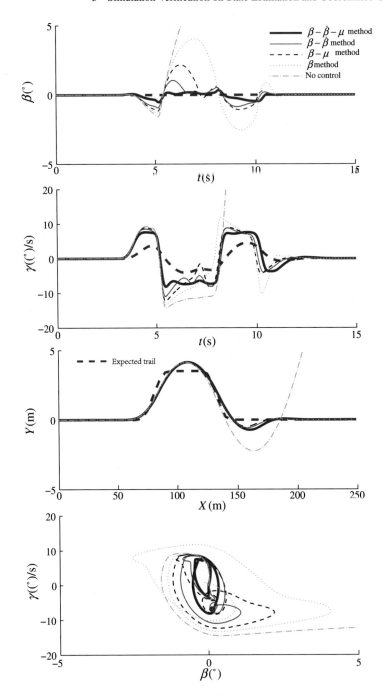

Fig. 5.28 Simulation of lateral stability control: double lanes

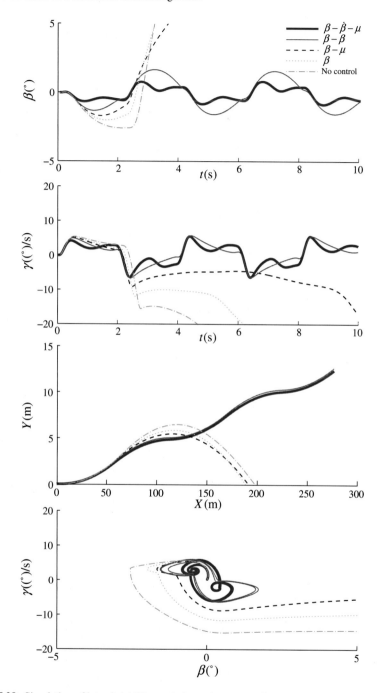

Fig. 5.29 Simulation of lateral stability control: continuous steering

Table 5.21 Simulation results of lateral stability control

No	Abbreviation	Double-lane			Continuous steering	
		max $|\beta|(°)$	max $|\gamma|(°/s)$	max $|\Delta Y|$(m)	max $|\beta|(°)$	max $|\gamma|(°/s)$
1	β method	4.05	13.31	0.6	–	–
2	$\beta - \mu$ method	2.17	12.35	0.6	–	–
3	$\beta - \dot{\beta}$ method	1.06	10.94	0.6	1.69	6.58
4	$\beta - \dot{\beta} -$ μ method	0.52	8.11	0.7	0.84	6.26

the five control schemes shown in Table 5.20. The definition of curves in Fig. 5.29 is consistent with that in Fig. 5.28. If direct yaw moment control is not performed, the side slip angle without control will have entered the instable area after 6 s. However, as the vehicle velocity is high, and the road surface adhesion coefficient is decreased, the direct yaw moment by "$\beta - \mu$ method" and the "β method" cannot ensure the stability of vehicle. After comparing the "$\beta - \dot{\beta}$ method" and "$\beta - \dot{\beta} - \mu$ method", the results will show that both of methods can control the vehicle within the stable area, and the maximum side slip angle in the "$\beta - \dot{\beta} - \mu$ method" control process is only 0.84°, which is less than 1.69° in the " $\beta - \dot{\beta}$ method", with more optimal control effect.

To simultaneously compare the control effects of different control methods, the maximum side slip angle max $|\beta|$, maximum yaw rate max $|\gamma|$ and lateral displacement deviation max $|\Delta Y|$ of different methods are compared. Refer to Table 5.21 for the simulation results of lateral stability control. As shown in Table 5.21, when the lateral stability control target is designed, if only the influence of size of side slip angle is considered, the control overshoot of side slip angle β will be obviously increased; if only the influence of side slip angle velocity $\dot{\beta}$ and road surface adhesion coefficient μ is considered, the control effect will be certainly improved. After comprehensively considering β $\dot{\beta}$ and μ, the control effect can be further optimized. Under the working conditions of double lane change and continuous steering, the "$\beta - \dot{\beta} - \mu$ method" can effectively reduce the side slip angle of vehicle, and ensure the lateral stability of vehicle.

5.4 Brief Summary

To verify the state estimation method and coordinated control method of distributed electric drive vehicle, this chapter developed the combined simulation platform through CarSim and Simulink. Basing on this, the state estimation method and coordinated control method were sequentially verified. The simulation results showed that the state estimation method of distributed electric drive vehicle can effectively observe multiple state parameters of full vehicle, i.e. mass, gradient, vertical force, lateral force, longitudinal velocity, side slip angle, yaw rate, etc., the estimation precision and real-time property are high, thus can be directly applied to the vehicle dynamic control system. Furthermore, the coordinated control method was verified by layers through simulation. Each subsystem can independently perform and organically combine various functions, improve the driving capability of full vehicle, and enhance the lateral stability of vehicle; under the conditions of drive trackslip and drive failure, effective and safe driving of vehicle can be guaranteed, and the advantages and huge potential of distributed electric drive vehicle can be comprehensively presented.

References

1. Shiozawa Y, Yokote M, Nawano M et al (2007) Development of a method for controlling unstable vehicle behavior. SAE technical paper: 2007–01–0840
2. Kinjawadekar T, Dixit N, Heydinger GJ et al (2009) Vehicle dynamics modeling and validation of the 2003 Ford expedition with ESC using CarSim. SAE technical paper: 2009–01–0452
3. Toyohira T, Nakamura K, Tanno Y (2010) The validity of EPS control system development using HILS. SAE technical paper: 2010–01–0008
4. Wilkinson J, Mousseau CW, Klingler T (2010) Brake response time measurement for a HIL vehicle dynamics simulator. SAE technical paper: 2010–01–0079
5. Bertollini G, Brainer L, Chestnut JA et al (2010) General motors driving simulator and applications to Human Machine Interface (HMI) development. SAE technical paper: 2010–01–1037
6. MacAdam CC (1980) An optimal preview control for linear systems. J Dyn Syst Meas Control 102(3):188–190
7. MacAdam CC (1981) Application of an optimal preview control for simulation of closed-loop automobile driving. IEEE Trans Syst Man Cybern 11(6):393–399
8. Pacejka HB (2012) Tire and vehicle dynamics, 3rd edn. Elsevier, Oxford

Chapter 6
Experimental Verification of State Estimation and Coordinated Control

Abstract To verify the effectivity of state estimation and coordinated control system for distributed electric vehicles, the existing experiment platform of distributed electric vehicles of laboratory is utilized for experiment verification. The state estimation experiments on the experiment platform mainly include experiment of mass estimation based on high-frequency information extraction, experiment of gradient estimation based on multi-method fusion, and experiment of vehicle motion state estimation based on unscented particle filter. The experiments of coordinated control mainly include experiment of motor property compensation control, experiment of driving force control allocation, and experiment of vehicle dynamic demand target determination. The experiments have verified the effectivity of methods proposed by this section, and the design purpose has been realized.

6.1 Experiment Platform of Distributed Electric Vehicle

The utilized experiment platform is the experiment distributed electric vehicle developed by Tsinghua group [1, 2], with the appearance of full vehicle as shown in Fig. 6.1. Refer to Table 6.1 for the main parameters of existing experiment platform of distributed electric vehicles. The software and hardware integration system of experiment vehicle mainly includes four parts, i.e. sensor system, control system, execution system and calibration system, which will be introduced as follows.

(1) Sensor system
It mainly utilizes the in-vehicle sensor and distributed driving motor to feedback the relevant information of vehicle. INS sensor integration system includes longitudinal acceleration sensor, lateral acceleration sensor and yaw rate sensor, with the sampling frequency of 100 Hz. GPS can collect the velocity and direction of vehicle, with sampling frequency of 1 Hz. The driver operation information collection system includes pedal opening degree sensor and steering angle sensor for collecting the information on pedal reference and steering angle, with sampling frequency of 100 Hz. The distributed driving motor feedback information system can provide the moment and rotation velocity of driving wheels in real time, with sampling frequency of 100 Hz.

© Springer-Verlag Berlin Heidelberg 2016
W. Chu, *State Estimation and Coordinated Control for Distributed Electric Vehicles*, Springer Theses, DOI 10.1007/978-3-662-48708-2_6

Fig. 6.1 Experiment Platform of distributed electric vehicles

Table 6.1 Main parameters of experiment platform of distributed electric vehicles

Parameters	Variable (unit)	Value
Gross weight	m (t)	2.2
Wheelbase	l (m)	2.3
Tread	b (m)	1.6
Tire rolling radius	R (m)	0.36
Motor power	P (kW)	15.7
Max torque of distributed driving wheel	T (Nm)	546

(2) Control system

It mainly utilizes the proposed state estimation and coordinated control algorithm, adopts the RCP developed by Matlab/Simulink, and loads the RCP into the in-vehicle dSPACE controller when the experiment is started. The dSPACE can complete the state estimation and coordinated control in real time, at the frequency of 100 Hz.

(3) Execution system

It is mainly used for executing the command of controller and command of drivers. The distributed driving motor is mainly used for executing the driving moment commands of each wheel issued by the coordinated control system. The steering system and the braking system directly respond to the operation of the drivers via mechanical connection.

(4) Calibration system
It mainly utilizes the vehicle motion state parameters directly observed by experiment equipment. To accurately obtain the vehicle motion state parameters, the RT3100 inertia navigation measurement system produced by Oxford Technical Solution Company is installed, and the measured value of RT3100 system is taken as the reference true value. The motion state parameters observed by the observation method are compared with the reference true value, so as to verify the effectivity of proposed motion state estimation algorithm.

6.2 Experiment of State Estimation System

The experiment is capable of verifying the effectivity and practicality of state estimation algorithm in the actual vehicle driving environment, and the proposed state estimation algorithm is verified on the experiment platform of distributed electric vehicles, mainly including (1) experiment of mass estimation based on high-frequency information extraction; (2) experiment of gradient estimation based on multi-method fusion; (3) experiment of vehicle motion state estimation based on unscented particle filter.

6.2.1 Experiment of Mass Estimation Based on High-Frequency Information Extraction

When the algorithm of mass estimation based on high-frequency information extraction is verified by experiment, it is necessary to consider the mass estimation results under different mass conditions, the influence of road surface gradient on mass estimation results, and the influence of different initial values of mass estimation on estimation results. Accordingly, three experiment working conditions are designed, as shown in Table 6.2.

Figure 6.2 shows the experiment results of mass estimation on flat road, wherein the initial value of mass estimation is 1t, and the actual mass is 2.20 t. The figure respectively shows the resultant longitudinal driving force, driving acceleration, and estimated value and true value of mass. When the estimation is started, the value

Table 6.2 Experiment working condition of mass estimation based on high frequency information extraction

No	Working condition	Vehicle mass (t)	Mass estimated initial (t)
1	Flat	2.20	1.00
2	Flat	2.12	3.00
3	Slope	2.20	1.00

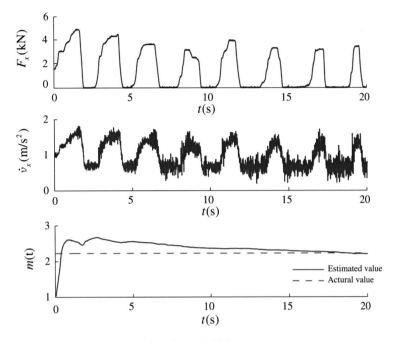

Fig. 6.2 Experiment of mass estimation: flat road, 2.20 t

is gradually convergent; after 16.1 s, the estimation error is convergent to the value within 3 % of true value, and the average value of mass estimation within 15–20 s is 2.23 t.

Figure 6.3 shows the experiment results of mass estimation on flat road, wherein the initial value of mass estimation is 3 t, and the actual mass is 2.12 t. The figure respectively shows the resultant longitudinal driving force, driving acceleration, and estimated value and true value of mass. When the estimation is started, the value is gradually convergent; after 17.1 s, the estimation error is convergent to the value within 3 % of true value, and the average value of mass estimation within 15–20 s is 2.17 t.

Figure 6.4 shows the experiment results of mass estimation on slope road, wherein the initial value of mass estimation is 1 t, and the actual mass is 2.20 t. The figure respectively shows the resultant longitudinal driving force, driving acceleration, and estimated value and true value of mass in the mass estimation process from top to bottom. When the estimation is started, the value is gradually convergent; after 13.8 s, the estimation error is convergent to the value within 3 % of true value, and the average value of mass estimation within 15–20 s is 2.24 t.

Refer to Table 6.3 for the summary of mass estimation results. According to Table 6.3, the proposed mass estimation algorithm can effectively estimate the mass of full vehicle within shorter time (about 20 s), and the error is convergent to the value within 3 %. After comparing the experiment 1 and experiment 2, the results

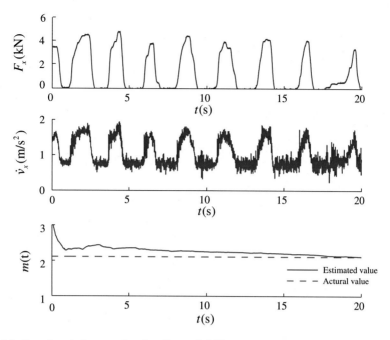

Fig. 6.3 Experiment of mass estimation: flat road, 2.12 t

show that when the mass of full vehicle is slightly changed (80 kg), the mass of full vehicle can be effectively estimated with the proposed estimation algorithm. After comparing the experiment 1 and experiment 3, the results show that under the condition of slope road, the proposed estimation algorithm is free from the influence of gradient, and the accurate estimated value of mass can be obtained within the shorter time. In experiment 1 and experiment 3, the initial value of mass estimation is 1 t; in experiment 2, the initial value of mass estimation is 3 t; the difference between initial value of mass and true value is large. However, the proposed algorithm still can be regressed to the true value within the shorter time, and is basically free from the influence of initial value, which indicates that the mass estimation algorithm is insensitive to the initial value of mass estimation and has better robustness.

6.2.2 Experiment of Gradient Estimation Based on Multi-method Fusion

The experiment of gradient estimation is performed on the basis of obtained mass estimated by the method in Sect. 3.1. In the experiment process, a section of uphill road is used, and when the vehicle velocity reaches about 5 km/h, the vehicle enters the uphill working condition, and starts to accelerate; after the experiment is fin-

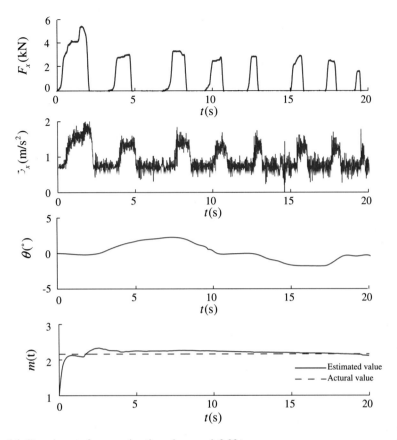

Fig. 6.4 Experiment of mass estimation: slope road, 2.20 t

Table 6.3 Experiment results of mass estimation based on high-frequency information extraction

No	Working condition	Real value (t)	Estimated value (t)	Error (%)	Convergency time (s)
1	Flat	2.20	2.23	1.4	16.1
2	Flat	2.12	2.17	2.4	17.2
3	Slope	2.20	2.24	1.8	13.8

ished, the longitudinal velocity of vehicle is about 20 km/h. Refer to Fig. 6.5 for the experiment results. The figure respectively shows the longitudinal driving force, driving acceleration, and estimated value and true value of mass in the mass estimation process from top to bottom, wherein the true value of gradient is measured by RT3100.

According to Fig. 6.5, the gradient is basically kept constant within 0–5 s; at this moment, the algorithm can well estimate the current gradient, and the gradient is gradually convergent. At 15–17 s, the road surface gradient is decreased suddenly,

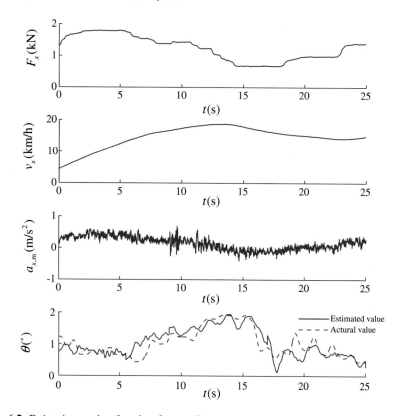

Fig. 6.5 Estimation results of road surface gradient

the estimated gradient value can follow the quick change of true value, leading to rapid response speed and high precision. In the whole estimation process, the maximum gradient error is 0.65°, and the RMS error is 0.11°. The experiment can verify the effectivity of proposed estimation algorithm, and the error of gradient estimation basically meets the actual requirement of real vehicle.

6.2.3 Experiment of Vehicle Motion State Estimation Based on Unscented Particle Filter

To verify the effectivity of vehicle motion state estimation algorithm, the vehicle motion state collected by RT3100 is used as reference true value, and the estimated value is compared with the reference true value. Firstly, the vehicle motion state estimation precision under the condition of longitudinal motion is verified; then, the vehicle motion state estimation precision under the condition of severe lateral motion is verified.

Table 6.4 Experiment working conditions of vehicle longitudinal motion state estimation

No	Adhesion coefficient	Working condition
1	0.8	Straight driving, continuous acceleration and brake
2	0.2	Straight driving, acceleration slip regulation
3	0.2	Straight driving, braking anti lock

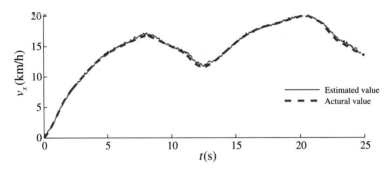

Fig. 6.6 Experiment of longitudinal velocity estimation: continuous acceleration and braking on road with high adhesion coefficient

(1) Longitudinal motion state observation

To verify the effectivity of velocity observation algorithm in the straight driving process of vehicle, the three working conditions in Table 6.4 are designed to observe the longitudinal velocity of vehicle.

Figure 6.6 shows the working condition of straight driving with continuous acceleration and braking on road with high adhesion coefficient. As the road surface adhesion coefficient is high, excessive slip does not occur on the vehicle. As the slip ration is low, and the in-wheel velocity is approximate to the vehicle velocity, the figure only compares the estimation vehicle velocity and reference vehicle velocity. The reference vehicle velocity is provided by RT3100 system. According to the observation results in Fig. 6.6, when the trackslip rate of vehicle is low, the RMS error of longitudinal velocity estimation is about 0.16 km/h.

Figure 6.7 shows the working condition of excessive slip on road with low adhesion coefficient. When the experiment is started, the driver controls the vehicle to start and accelerate from the idle state; as the road surface adhesion coefficient is low, excessive slip occurs to the right front wheel firstly at 7.2 s. When the motor is subject to overspeed, the motor controller will start the self protection mechanism, and the driving force command is disconnected; therefore, the in-wheel velocity of right front wheel will quickly be decreased to near vehicle velocity after about 1 s. At 7.5 s, as the driving force is continuously increased, and excessive slip will occur to all wheels; when the driver decreases the longitudinal driving force of vehicle, the

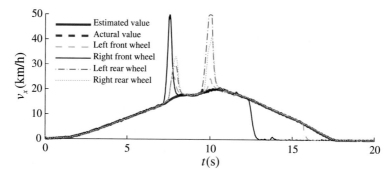

Fig. 6.7 Experiment of longitudinal velocity estimation: over slip on road with low adhesion coefficient

slip ratio of each wheel is decreased again, and the in-wheel velocity is regressed to near vehicle velocity. At 9.5–10.5 s, due to the severe operation of driver, tire over slip will occur to the vehicle again. As the road surface adhesion coefficient is low, brake locking respectively occurs to the right front wheel and left front wheel in the braking process at 12.3 and 15.6 s. In the whole drive and braking process, the drive trackslip and brake locking will occur for multiple times, and even the slip of all wheels occurs, the system still can accurately estimate the vehicle velocity, with the RMS error of longitudinal velocity under the single working condition being 0.19 km/h.

Figure 6.8 shows the working condition of braking slip on road with low adhesion coefficient. When the experiment is started, the vehicle is in the idle state. The vehicle starts and accelerates after 2.5 s, and as the desired longitudinal driving force of driver is small, excessive slip does not occur to the vehicle. At 11.4 s, the vehicle is subject to emergently braking, and the condition of braking slip occurs because of overlarge braking force and low road surface adhesion coefficient. At 0.5 s after emergency braking, the slip rate of front wheel is quickly increased; at 0.8 s after emergency

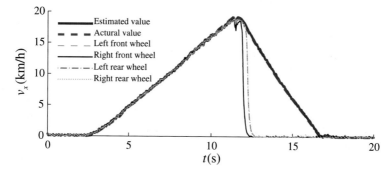

Fig. 6.8 Experiment of longitudinal velocity estimation: braking on road with low adhesion coefficient

Table 6.5 Experiment results of vehicle longitudinal motion state estimation

No	Working condition	RMS error (km/h)	Max error (km/h)
1	Continuous acceleration and brake at high adhesion coefficient	0.16	1.12
2	Acceleration at low adhesion coefficient	0.19	1.24
3	Brake at low adhesion coefficient	0.22	1.83

braking, the slip rate of rear wheel is quickly increased, and the period from quick increase of slip rate to complete locking of wheel is only 0.8 s. Subsequently, the vehicle is always kept in full-locking state until completely stopping. The RMS error of longitudinal velocity under the single working condition is 0.22 km/h.

Table 6.5 shows the experiment results of vehicle longitudinal motion state estimation. Under the working condition of continuous acceleration and deceleration on road with high adhesion coefficient, or excessive trackslip and slip on road with low adhesion efficient, the longitudinal velocity of vehicle can be effectively evaluated with observation algorithm, with the RMS error smaller than 0.3 km/h, and the maximum error smaller than 2 km/h.

(2) Lateral motion state observation

To observe the motion state of vehicle under the condition of severer lateral motion, the experiment selects the working conditions of double lane change and continuous steering for combined observing the longitudinal velocity, yaw rate and side slip angle of vehicle, and considers that the vehicle motion state obtained by RT300 system is the true value. Under the working condition of double lane change, the vehicle motion state estimation results are shown in Fig. 6.9. The figure respectively shows the estimated value and true value of longitudinal velocity, yaw rate and side slip angle. When the experiment is started, the vehicle accelerates along a straight line. When the longitudinal velocity of vehicle reaches about 30 km/h, the driver controls the vehicle to move under the condition of double lane change. In this process, the longitudinal velocity of vehicle is about 30 km/h. The maximum error of longitudinal velocity in the whole estimation process is 0.77 km/h, and the RMS error is 0.26 km/h. The maximum error of side slip angle occurs at about 6 s, with the maximum error of 0.67° and RMS error of 0.24°. The maximum error of yaw rate is 5.24°/s, and the RMS error is 1.42°/s.

Under the condition of continuous steering, the vehicle motion state estimation results are shown in Fig. 6.10. When the experiment is started, the vehicle velocity is about 30 km/h; subsequently, the vehicle is accelerated to about 35 km/h in a linear way, with the velocity maintained. The driver controls the vehicle for continuous steering. The maximum error of longitudinal velocity in the whole estimation process is 1.01 km/h, and the RMS error is 0.47 km/h. The maximum error of side slip angle in the whole estimation process is 0.55°, and the RMS error is 0.21°. The maximum

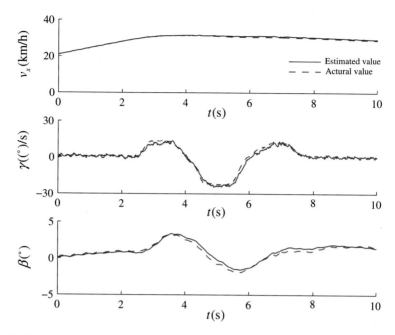

Fig. 6.9 Experiment of lateral motion state observation: double lanes

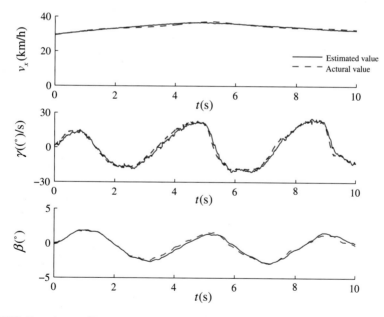

Fig. 6.10 Experiment of lateral motion state observation: continuous steering

Table 6.6 Experiment results of vehicle lateral motion state estimation

No	Working condition	RMS error			Max error		
		v_x (km/h)	γ (°/s)	β (°)	v_x (km/h)	γ (°/s)	β (°)
1	Double lanes	0.26	1.42	0.24	0.77	5.24	0.67
2	Continuous steering	0.47	1.95	0.21	1.01	6.78	0.55

error of yaw rate in the whole estimation process is 6.78°/s, and the RMS error is 1.95°/s.

Refer to Table 6.6 for the two groups of observation results. After analysis, under the conditions of double lane change and continuous steering, the vehicle is in the strong non-linear observation motion phase, and the observation precision of longitudinal velocity, side slip angle and yaw rate is high, which can effectively reflect the real motion state of vehicle.

6.3 Experiment of Coordinated Control

The experiment is capable of verifying the effectivity and practicality of coordinated control algorithm in actual vehicle driving condition, and the proposed coordinated control algorithm is verified on the experiment platform of distributed electric vehicles, mainly including (1) experiment of motor property compensation control; (2) experiment of coordinated control; (3) experiment of vehicle dynamic demand target determination.

6.3.1 Experiment of Motor Property Compensation Control

To verify the effectivity of motor property compensation control algorithm, it is necessary to consider the input moment commands under different properties, and the effects of motor property compensation control. When the motor property compensation control is performed, the difference between motor feedback moment and expected moment under the conditions of steady-state ascending and continuous change of input command is considered. The motor actual output moment is provided by the motor controller. The designed experiment working conditions are shown in Table 6.7.

To compare the driving moment responses under the working conditions with or without motor property compensation control, the expected output moment and actual output moment are compared (under the working conditions with or without control). To compare the working conditions with or without control in one experiment, the moment responses of two motors are simultaneously compared, wherein

Table 6.7 Experiment working conditions of motor property compensation control

No	Input torque command	Error exist
1	Steady increase	Static error, ratio error, dynamic error
2	Continuous changes	Static error, ratio error, dynamic error

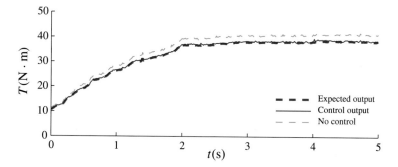

Fig. 6.11 Experiment of motor property compensation control: steady ascending

the moment of one motor is subject to property compensation control, while the moment of the other motor is subject to open-loop control.

Refer to Fig. 6.11 for the experiment of property compensation control under the condition of steady-state ascending of motor moment. The figure compares the expected output moment, compensation control output moment and non-compensation output moment. There is larger difference between output moment value under non-compensation control condition and expected output moment value, with the average value of about 2.75 N m. After the motor property compensation control is performed, the average error of output moment is decreased to 0.19 N m.

Refer to Fig. 6.12 for the experiment of property compensation control under the condition of continuous steering of motor moment. The figure compares the expected output moment, compensation control output moment and non-compensation

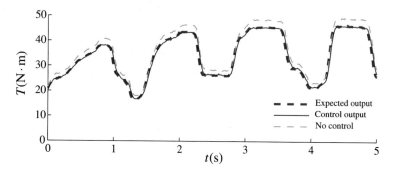

Fig. 6.12 Experiment of motor property compensation control: continuous change

Table 6.8 Experiment results of motor property compensation control

No	Working condition	No control error (N m)	Control error exist (N m)
1	Steady increase	2.75	0.19
2	Continuous changes	2.40	0.11

output moment. The average error between the output moment value under non-compensation control condition and expected output moment value is 2.40 N m. After the motor property compensation control is performed, the average error of output moment is decreased to 0.11 N m.

Under different input commands, refer to Table 6.8 for the average difference between output moment and expected moment under the working conditions with or without motor property compensation control. When the property compensation control is not performed, the error of motor output moment is about 7 % of expected moment; after the property compensation control is performed, the average error of output moment is about 0.5 % of expected moment. After the property compensation control, the control precision of moment is greatly improved. The property compensation control can effectively improve the dynamic control precision of full vehicle, and improve the driving property of vehicle.

6.3.2 Experiment of Control Allocation

The experiment of driving force control allocation is mainly used for verifying the driving force control allocation condition under the working conditions of drive trackslip and drive failure. The three working conditions, i.e. driving force control allocation, non-coordinated control and normal driving (without drive failure and drive trackslip) are compared. Simultaneously, the dynamic property and stability are comprehensively considered, and the designed motion working conditions include low-velocity starting and high-velocity motion. Under various working conditions, the steering wheel angle open-loop operation is adopted, with the steering wheel angle being 0. To ensure the comparison and repetitiveness of experiments, the desired longitudinal driving force of driver is set as fixed value under various conditions, while the desired direct yaw moment is set as 0. Therefore, the expected longitudinal force of each wheel is an equal fixed value when the drive failure and drive trackslip do not occur.

In the low-velocity phase, the driving force control allocation under the conditions of drive failure and drive trackslip is mainly considered. According to Sect. 4.2.1, the desired longitudinal driving force is firstly met. Under the working condition of low-velocity trackslip, the vehicle starts and accelerates from the idle state on the bidirectional road, and the right wheels run on the road with low adhesion coefficient. Under the working condition of low-velocity failure, the vehicle starts and accelerates from the idle state on the road with high adhesion coefficient, when partial failure occurs to the right front wheel after 1 s and to left rear wheels after 3 s. In the high-velocity phase, the driving force control allocation under the condition of drive failure is mainly considered. According to Sect. 4.2.1, the desired direct yaw moment is firstly met. Under the working condition of high-velocity failure, partial failure occurs to the right front wheel 1 s after the vehicle velocity exceeds 20 km/h and to left rear wheel 3 s after the same.

Figure 6.13 shows the dynamic coordinated control results under the working condition of low-velocity starting. The figure compares the vehicle running states under the conditions of no failure, non-coordinated control and coordinated control, and sequentially compares the longitudinal driving force, direct yaw moment, longitudinal velocity, motion track, and driving force value of each wheel from top to bottom. In the experiment process, the vehicle starts and accelerates from the idle state, the desired longitudinal driving force is set as a fixed value, and the desired direct yaw moment is 0. Firstly, the working condition of low-velocity starting failure without coordinated control is analyzed. Partial failure occurs to the right front wheel 1 s after starting, and the generated maximum driving force is less than the expected moment value. Therefore, if the coordinated control is not performed, the desired longitudinal driving force of full vehicle will not be met. Partial failure occurs to the left rear wheel 3 s after starting, and the generated maximum driving force is decreased to 20 % of original value. The right front motor and left rear motor cannot execute the expected driving force after 3 s; if the coordinated control is not performed, the dynamic performance of full vehicle will further decrease. After 5 s, the longitudinal velocity of vehicle is 14.6 km/h. Then, the working condition of coordinated control is analyzed. After partial failure occurs to the right front wheel, the control system will coordinate and increase the longitudinal driving forces of the other three driving wheels, and the longitudinal driving force requirement of full vehicle can be met within 1–3 s. After 3 s, as the two wheels fail, the system controls all wheels to work at full load, and the driving forces of all wheels have been saturated; although the driving force is slightly lower than the expected longitudinal driving force, it is the optimum force under such condition. After 5 s, the longitudinal velocity of vehicle is 16.7 km/h. Under the condition of no failure in low-velocity starting, after 5 s, the longitudinal velocity of vehicle is 18.8 km/h. The comparison results show that the coordinated control algorithm based on dynamic property can effectively improve the starting and acceleration capability of vehicle in the low-velocity staring phase.

Figure 6.14 shows the dynamic coordinated control results under the working condition of trackslip in low-velocity starting. If the drive trackslip control is not performed, excessive trackslip quickly occurs to the wheels within shorter time, when the acceleration test will not be able to be performed. The figure only compares the vehicle running states under the conditions of no trackslip, non-coordinated control and coordinated control, but does not compare the working condition of drive trackslip control. Refer to essays of the author for control effect of single wheel drive trackslip [3]. The definition of curves in Fig. 6.14 is consistent with that in Fig. 6.13. In the experiment process, the vehicle starts and accelerates from the idle state, the desired longitudinal driving force is set as fixed value, and the desired direct yaw moment is 0. Firstly, the working condition of low-velocity starting trackslip with-

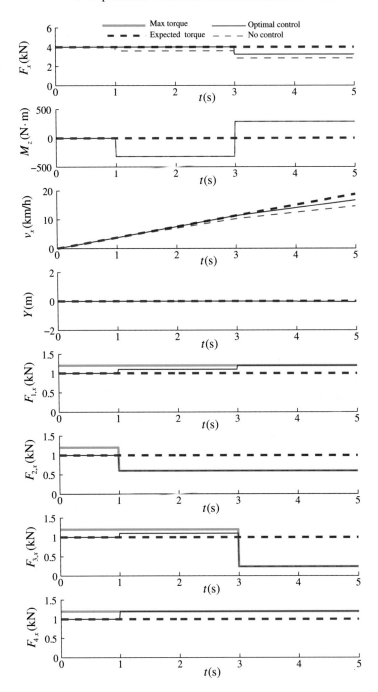

Fig. 6.13 Experiment of coordinated control: fail in low speed start

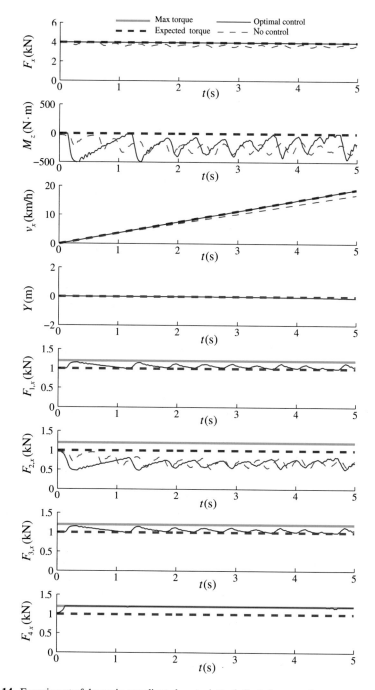

Fig. 6.14 Experiment of dynamic coordinated control: trackslip in low speed start

out coordinated control is analyzed. Partial drive trackslip occurs to the right front wheel immediately after starting, when the TCS is adopted to decrease the driving force command of right front wheel, for the wheel cannot continue to execute the expected driving force. When the coordinated control is not performed, the expected driving force cannot be executed, and the dynamic performance of full vehicle will be greatly influenced. After 5 s, the longitudinal velocity of vehicle is 17.0 km/h; if the coordinated control is adopted, and the TCS is adopted to control, the driving force of the other three driving wheels can be coordinated to ensure the execution of expected longitudinal force. To avoid overlarge direct yaw moment, the driving force of right rear wheel is firstly increased. After 5 s, the longitudinal velocity of vehicle is 18.8 km/h. Under the condition of no trackslip in low-velocity starting, the longitudinal velocity of vehicle is also 18.8 km/h after 5 s, and the acceleration capability under coordinated control condition keeps consistent with that of the vehicle without trackslip. The comparison results show that the coordinated control algorithm based on dynamic property can effectively improve the starting and acceleration capability of vehicle in the low-velocity starting phase under the condition of trackslip.

Figure 6.15 shows the stability coordinated control results under the working condition of failure in high-velocity motion. The figure compares the vehicle running states under the conditions of no failure, non-coordinated control and coordinated control. The definition of curves is consistent with that in Fig. 6.13. In the experiment process, the vehicle starts and accelerates from the idle state, the desired longitudinal driving force is set as a fixed value, and the desired direct yaw moment is 0. Partial failure occurs to the right front wheel 1 s after the vehicle velocity exceeds 20 km/h, and the wheel cannot continue to execute the expected driving force. When the coordinated control is not performed, as the driving force at right side is suddenly decreased, unexpected direct yaw moment will be generated, and the lateral stability of full vehicle will be greatly influenced. After 5 s, the lateral displacement of vehicle is 1.1 m, which is very dangerous. If the driving wheels of vehicle fail at high velocity and timely coordinated control is not performed, severe consequence will be caused. If the coordinated control is performed, the driving force of the other three driving wheels can be coordinated to avoid unexpected direct yaw moment. After 5 s, large lateral displacement does not occur to the vehicle. It can be seen by comparing the lateral displacements that, after the coordinated control is performed, safety of full vehicle is ensured.

Refer to Table 6.9 for all experiment results. It is noticeable that as the lateral displacement <0.1 m is of less significance, it can be regarded as 0. In the low-velocity starting phase, according to the proposed optimum control method, the longitudinal driving force can be effectively ensured under the conditions of failure (working condition 2) and trackslip (working condition 4), the starting is quick, and the starting and acceleration property is obviously superior to those without coordinated control (working condition 3 and working condition 5). Although certain unexpected yaw moment is generated (target direct yaw moment of 0), as the vehicle velocity is low, large lateral displacement does not occur to the vehicle, so stability can be guaranteed to a certain extent. Compared with the normal driving condition (working condition 1), after the drive failure and drive trackslip occur, although the longitu-

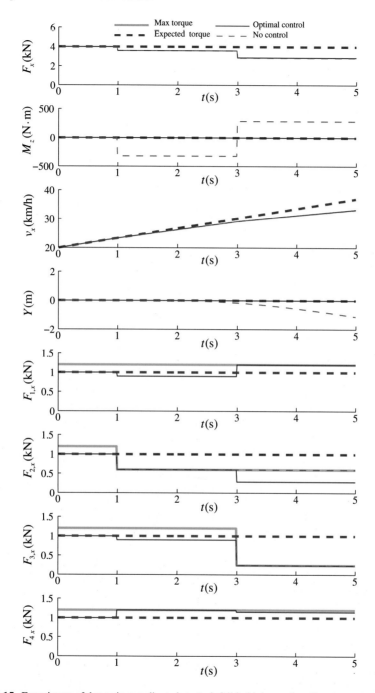

Fig. 6.15 Experiment of dynamic coordinated control: fail in high speed motion

Table 6.9 Working conditions and results of driving force control allocation

No	Working condition	Failure	Slip	Control coordination	End speed (km/h)	Lateral disp (m)
1	Start at low speed	No failure	No slip	Average distribution	18.8	0
2	Start at low speed	Fail	No slip	Dynamic coordination	16.7	−0.1
3	Start at low speed	Fail	No slip	No coordination	14.6	−0.1
4	Start at low speed	No failure	Slip	Dynamic coordination	18.8	−0.1
5	Start at low speed	No failure	Slip	No coordination	17.0	−0.1
6	Start at high speed	No failure	No slip	Average distribution	37.0	0
7	Start at high speed	Fail	No slip	Steady coordination	33.2	0
8	Start at high speed	Fail	No slip	No coordination	33.2	−1.1

dinal driving capability of part of wheels suffers big loss, the overall longitudinal driving capability of vehicle can be better maintained through coordinated control of each wheel. In the high-velocity motion phase, the lateral stability of vehicle can be effectively guaranteed under the condition of failure (working condition 7) by the optimum control method, unexpected direct yaw moment is reduced, and the yaw of vehicle in the straight driving process is further reduced. Compared with the condition without coordinated control (working condition 8), certain longitudinal driving force is maintained, the yaw and lateral displacement are reduced, so the safety of vehicle is guaranteed. In general, the coordinated control algorithm can effectively improve the starting and acceleration capability of vehicle in the low-velocity phase, and ensure the stability of vehicle in the high-velocity phase.

6.3.3 Experiment of Vehicle Dynamic Demand Target Determination

The experiment of vehicle dynamic demand target determination is mainly used to verify the control target of vehicle, especially about whether the control target of direct yaw moment can be effectively executed or not, and whether the target can ensure the stability of vehicle or not. Accordingly, three groups of experiments are designed, as shown in Table 6.10. Under the working conditions of double lane change and continuous steering, partial failure has occurred to the two front wheels.

Table 6.10 Experiment of vehicle dynamic demand target

No	Longitudinal control	Lateral control	Failure working condition
1	Constant speed	Closed loop, double lane change	Failure of front wheel
2	Constant speed	Open loop, continuous steering	Failure of front wheel
3	Constant speed	Closed loop, circular motion	No failure

Refer to Figs. 6.16 and 6.17 for the experiment results of working conditions of double lane change and continuous steering. To directly proof the effect of set lateral control target, the circumference motion experiment of fixed steering wheel angle is designed. In the experiment process, each driving wheel can work normally. Refer to Fig. 6.18 for the experiment results.

Figure 6.16 shows the lateral control results under the working condition of double lane change. The figure sequentially shows the resultant longitudinal driving force, direct yaw moment, side slip angle, expected value and true value of yaw rate, and longitudinal driving force of each wheel. The vehicle can be controlled within the stable area by the proposed direct yaw moment control scheme, with the maximum value of side slip angle being about 1.4°. Meanwhile, the actual yaw rate can well follow the expected yaw rate. Under the condition of partial failure in front wheels, the control system can optimize the motor utilization rate to protect the failed motor, with the distributed driving force of rear wheel being greater than that of front wheel.

Figure 6.17 shows the lateral control results under the working condition of continuous steering. The definition of curves in the figure is consistent with that in Fig. 6.16. The vehicle is controlled within the stable area by the proposed direct yaw moment control scheme, with the maximum value of side slip angle being about 2.0°. Likewise, the actual yaw rate can well follow the expected yaw rate. The distributed driving force of rear wheel is greater than that of front wheel, so the motor utilization rate is optimized and the failed motor is protected. Therefore, under the conditions of double lane change and continuous steering, the set vehicle dynamic demand target can control the side slip angle within the reasonable area, and the expected yaw rate can be effectively followed. In the driving force control allocation process, the distributed driving force of normal wheels is larger than that of failure vehicles, so the motor utilization rate is optimized and the failed motor is protected.

Refer to Fig. 6.18 for the experiment results of circumference motion. In the experiment process, the steering wheel angle and acceleration pedal opening degree are fixed, and the vehicle performs circumference motion. Meanwhile, a direct yaw moment control switch is designed: when it is switched on, the set direct yaw moment is executed; when it is switched off, the control of direct yaw moment is not performed. In the experiment process, the change in yaw rate is mainly compared, wherein the direct yaw moment control switch is on at 2 s. According to Fig. 6.18, the direct yaw moment control is not performed before 2 s, so there is difference between actual yaw rate and target yaw rate of vehicle, being about 4°. After 2 s,

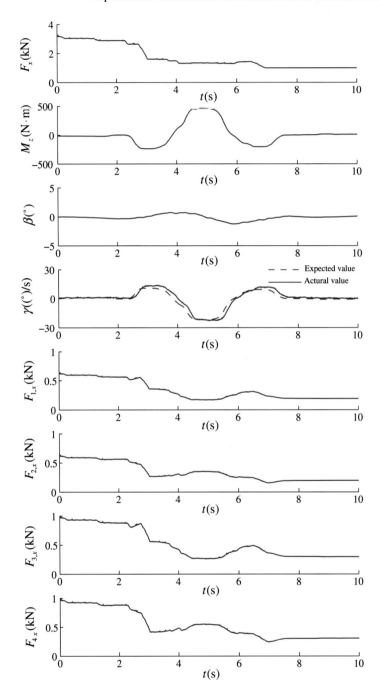

Fig. 6.16 Experiment of vehicle demand target determination: double lanes

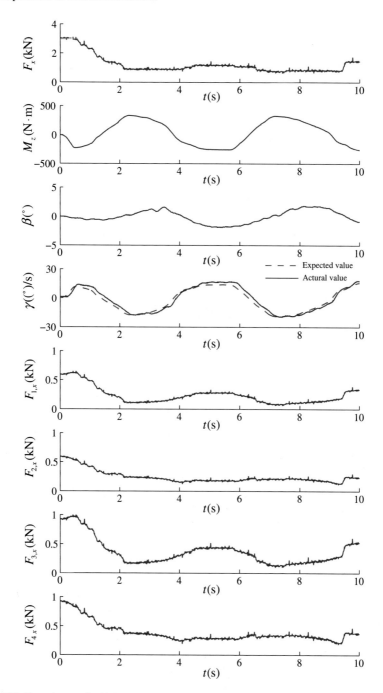

Fig. 6.17 Experiment of vehicle demand target determination: continuous steering

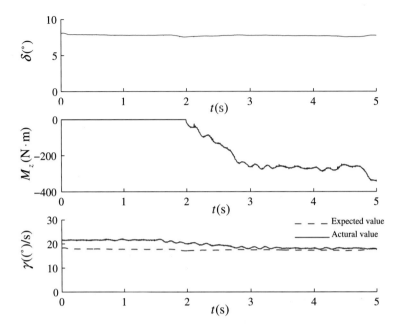

Fig. 6.18 Experiment of vehicle demand target determination: circumference motion

as the direct yaw moment control switch is on, the actual yaw rate is convergent to near the target yaw rate after about 1 s, which directly proofs the control effect of the proposed lateral control object.

6.4 Brief Summary

To verify the effectivity of state estimation and coordinated control system for distributed electric vehicles, experiment verification was performed on the existing experiment platform of distributed electric vehicles of laboratory. The experiment results showed that the proposed state estimation method of distributed electric vehicles can effectively observe multiple state parameters of full vehicle, i.e. longitudinal velocity, side slip angle, yaw rate, mass of full vehicle, road surface gradient, etc., with satisfactory estimation precision and real-time property. In the coordinated control process, the experiment verified the effectives of motor property compensation control algorithm, driving force control allocation algorithm and vehicle dynamic demand target determination algorithm through comparison. The experiment showed that the proposed coordinated control method can comprehensively guarantee the stability and driving capability of vehicle, thus playing an important role in enhancing dynamic potential of full vehicle.

References

1. Wang B, Luo Y, Zou G et al (2009) Experiment platform for study on four-wheel independently-driven off-road vehicle. J Tsinghua Univ (Nat Sci) 49(11):1838–1842
2. Wang B (2009) Study on experiment platform and driving force control system for four-wheel independently-driven electric vehicle. Tsinghua University, Beijing
3. Chu W, Luo Y, Zhao F et al (2012) Driving force coordinated control of distributed electric drive vehicle. J Automot Eng 34(3):185–189

Chapter 7
Conclusions

To improve the dynamic performance of distributed electric drive vehicle, the features of driving motor independent control and multi-information fusion of distributed electric drive vehicle were utilized, the study on the state estimation and coordinated control of distributed electric drive vehicle were completed, and the correctness and effectivity of method were verified by the CarSim and Simulink combined simulation and actual vehicle experiment. The conclusion is as follows:

(1) By virtue of the features of driving motor independent control and multi-information fusion of distributed electric drive vehicle, the state estimation and coordinated control system of distributed electric drive vehicle is designed, which optimizes the dynamic control property of full vehicle, and improves the stability, safety and driving capability of vehicle.

(2) The algorithm of mass estimation based on high-frequency information extraction can estimate the mass only by utilizing the information on longitudinal driving force and driving acceleration, which avoids the influence of road surface gradient on mass estimation results, and improves the mass estimation precision.

(3) The algorithm of gradient estimation based on multi-method fusion integrates the high-frequency information of kinematic observation results and low-frequency information of dynamic observation results, which simultaneously improves the observation precision of slow-varying gradient and quick-varying gradient.

(4) The algorithm of vertical force estimation based on multi-method fusion can observe the lateral force of tire with the estimated roll angle and roll angle velocity, which improves the observation precision of lateral force.

(5) The algorithm of vehicle motion state and lateral force based on unscented particle filter adopts the unscented particle filter method to jointly observe the longitudinal velocity, side slip angle, yaw rate and wheel lateral force of full vehicle, which improves the vehicle state estimation precision under strong non-linear motion state.

© Springer-Verlag Berlin Heidelberg 2016
W. Chu, *State Estimation and Coordinated Control for Distributed Electric Vehicles*, Springer Theses, DOI 10.1007/978-3-662-48708-2_7

(6) The algorithm of vehicle dynamic demand target determination adopts the $\beta - \dot{\beta}$ phase diagram to estimate the driver command and vehicle motion state, and sets the direct yaw moment meeting the driver command and stability requirement, which improves the control effect of lateral motion and vehicle safety.

(7) The algorithm of driving force control allocation comprehensively considers many constraint conditions, effectively ensures the driving capability of vehicle under the conditions of drive trackslip and drive failure, and maintains the lateral stability of vehicle; when there is a contradiction between the desired longitudinal driving force and desired direct yaw moment, the weight of desired longitudinal and lateral demand will be subject to real-time coordination to firstly met the main demand target, and therefore the driving force control allocation problem of distributed electric drive vehicle is completely solved.

(8) The algorithm of motor property compensation control utilizes the Lyapunov method to design the motor property compensation control method based on self-adaptive control according to the different driving motor properties of distributed electric drive system, so as to solve the consistency problem of motor properties.

(9) With the combined simulation platform developed by CarSim and Simulink and the real vehicle experiment platform built by laboratory, it is verified that the proposed dynamic state observation can effectively estimate the state parameters of vehicle under multiple working conditions, and the proposed Coordinated Control system can coordinate and control the driving forces of multiple wheels, thus comprehensively optimizing the longitudinal and lateral dynamic performance of full vehicle.

There are following innovative aspects in this dissertation:

(1) The method for vehicle state parameter estimation based on unscented particle filter is proposed. The method utilizes the unscented particle filter to jointly observe multiple vehicle state parameters, combines the kinematic model and dynamic model of vehicle, integrates the wheel moment/rotation velocity information, GPS information, INS information and driver operation information, and solves the problem of reduced observation precision of state parameters in case of obvious non-linear motion property of vehicle.

(2) The method for driving force control allocation of distributed electric drive vehicle with many constraints is proposed. The method designs the constraint conditions and objective function, solves the driving force control allocation problem of distributed electric drive vehicle under the conditions of drive trackslip and drive failure, and furthest meets the longitudinal and lateral targets of full vehicle under the conditions of drive trackslip and drive failure. When there is a contradiction between longitudinal and lateral demand control targets, the method can comprehensively optimize and control the drivers' demand and current vehicle state. When drive failure occurs, the method can decrease the utilization rate of motor to protect the failed motor.

(3) The method for motor difference compensation control based on model matching is proposed. Specific to the difference in driving motor of existing distributed electric drive motor, the method utilizes the self-adaptive control theory to design the motor difference compensation control algorithm, and adopts the Lyapunov method to design the self-adaptive law, in order to ensure the stability of control. The method can compensate the steady-state properties and dynamic properties of motor in real time, standardize the input and output properties of multiple controlled motors, and ensure the consistency of steady-state properties and dynamic properties of each driving motor of distributed electric drive vehicle.

This dissertation has obtained the systematic achievements on State Estimation and Coordinated Control system for distributed electric drive vehicle, but there are still some subsequent achievements to be studied, as follows:

(1) The proposed state estimation method of distributed electric drive vehicle only focuses on several common state parameters in the dynamic field, but there are still many vehicle state parameters requiring to be observed, i.e. center of mass, yaw inertial, rolling radius of tire, air pressure of tire, etc. If these parameters can be observed, the vehicle dynamic control and effects of other vehicle controllers can be further improved. Therefore, it is necessary to further study the observation of multiple vehicle state parameters. Meanwhile, as the proposed State Estimation method is only suitable for distributed electric drive vehicle, it is necessary to discuss the application of method on traditional vehicles and other types of vehicles.

(2) The proposed failure control of distributed electric drive vehicle does not contain the failure diagnosis of electric driving wheels, and the failure information or moment limitation information of corresponding motor can only be obtained from the feedback information of motor controller. If the failure mode of driving wheels and failure condition of other parts can be effectively detected by virtue of the information on distributed electric drive vehicle itself, the application range of failure control will be further widened.